AF887990

JOHN KHOURY

QUANOLOGY

EVOLUTION & YOU

Copyright © 2014 by John Khoury

All rights reserved. No part of this publication may be reproduced, distributed, or transmitted in any form or by any means, including photocopying, recording, or other electronic or mechanical methods, without the prior written permission of the publisher, except in the case of brief quotations embodied in critical reviews and certain other noncommercial uses permitted by copyright law. For permission requests, write to the publisher, addressed "Attention: Permissions Coordinator," at the address below.

Quanology
www.quanology.org

ISBN: 0990856607
ISBN 13: 9780990856603
Library of Congress Control Number: 2014952238
Quanology, Towson, MD

DEDICATION

This one goes out to the one I love. Meanwhile, we've made a few more people I love. My journey of discovery began when I met you. I never knew people could be like you — so warm, friendly, caring, and selfless. And that body. . .

You've shown me how important a woman is to a man who wants so much out of life and makes himself crazy in the process - walking the line between obsessive work and sitting on the couch with his hand down his pants.

While I'm acutely aware of how little you will appreciate this book, all these years, you have believed in me, nurtured the good side of me, helped me to be a better person — more like you — and taught me to love, to be open, and to not be afraid of who I am.

As all these things come so naturally and effortlessly to you, I am not so much grateful as I am inspired, in awe, and in love.

CONTENTS

Dedication v

Introduction 1

1 | Quanology 3
A Very Good Place to Start 7
The Basics 11
The Star of the Show: Technology 15

2 | Evolution 2.0 17
Big Bang Baby 18
Evolution Is the Creator 19
You're So Gene-ish 24
The Sweet Spot 25
The Great Loophole 28

3 | Technology: The Secret Ingredient 31
Evolution & Math 32
A New Look at Technology 43
GPP: The Value of a Human Being 64
Technology & Culture 77
Life Strategies 108

4 | The Human Experience 135
Sex & Love 136
Men versus Women 143
Family 166
Government 171
Economics & Work 179
Education 189
Wars & Culture Wars 192
Violence 203
Drugs 208
Art 212
Charity 220
Progress 226
Digital Evolution 235

5 | Personal Development 253
About You 253
Your Purpose 259
Getting to Work: Evolution in Your life 272
The Principles of Evolution in Your Life and How to Use Them . 275

6 | The Meaning of Life 311

The End 315
Special Thanks 315
Notes 317

Introduction

> ***"Living is easy with eyes closed"***
> — John Lennon

I was 12 years old when I read those lyrics scribbled on the ceiling of a tent at summer camp. They were like a giant hand that grabbed me by the waist and pulled me into a new world. In those words were depth and meaning, sung by someone who understood and had the guts to tell the world it was screwing up. I had heard those lyrics before, but this time it was different: I was open to something new that allowed them to hit me in a profound way. All of a sudden, I felt a connection to someone or something that thought like I thought. I connected not only with their author, but also with the guy who wrote them on the tent ceiling. I realized that there was a world of people out there who were just as confused and dissatisfied as I was, and I had just joined their team.

But in that moment of emotional clarity, it was suddenly apparent that everything else in my life – school, my family, girls, my future – was unclear. Those words acknowledged that in this life, there are people who go through the motions, and there are others who step out of line and look for more. So, maybe there were more people like me, but where were they? I hadn't ever met them, but at least, through that song, I knew they were there. I couldn't see them, but through thoughts in the ether I was connected to them.

Many years later I realized those words were wrong – living is actually easier with your eyes open. John Lennon was being insulting, complaining about all those "straights" who went through life with blindfolds on, while he, poor guy, was unable to close his eyes. He was cursed with understanding. He saw the world with eyes wide open and it was hard. This book hopes to change that for you. I want to open your eyes and, in the process, make your life easier.

You may or may not be aware that we have reached a truly historic moment in the world. With murder, war, genocide, and rape protruding like unrelenting weeds in a garden of love, charity and kindness, life is a puzzle that we want to figure out. Despite this seeming paradox, there is a simple method to the madness in the world. More importantly, this method is relevant to your life. Ultimately what we want to know is: How do we get more good than bad out of our lives — for ourselves and for the whole world?

Quanology is here to answer that question. It is a framework for living based on the underlying current of life: evolution. Evolution is an all-powerful force that rules over everything in the living world. Once you understand evolution you'll understand life on a whole new level and connect with the world like you never have before. Nature is not a separate entity, living its own life alongside us. We are nature, too, and when we study nature we learn about ourselves. I'm going to spend the rest of this book sharing that understanding with you. The process won't be easy – I'm sure you'll be annoyed or offended at least a few times. But maybe what you learn here will grab you by the waist and pull you into a new world.

Quanology is that new world.

So, welcome. I'm here to give you a new blueprint for a successful existence in today's world and a new team to join that will provide you with a better sense of direction and focus in your pursuit of the best life possible.

1 | Quanology

Quanology is a look at life through the eyes of evolution. We all want a good life and having a good life starts with understanding what that life is all about. We need to go back to the ground floor: What are we doing here? What's our purpose? Evolution tells us that we're animals, and that means we need to get to know ourselves as such, before we can understand ourselves in this modern world.

When I was 21, I got contact lenses. Everything was clear all of a sudden. It was like "Holy shit, I can see!" The world looked brand new. Quanology provided me with a similar experience. Evolution is a new pair of glasses that allow us to take a fresh look at everything.

When we piece a few of the insights Quanology has to offer, we get a very compelling picture. The first piece is that we are animals, created by evolution over billions of years. Despite the modern aspects of your life, your animal instincts still comprise the basis for your actions. The next piece is seeing that when animals change over time, what they are really doing is developing technology. That new wing that an animal develops is a technology to help it do better in the world. Instead of developing new body parts, modern humans add cultural technologies like buying a hammer or getting a degree in engineering. But these modern technologies are still a part of nature – just as much as all the biological technologies that plants and animals are constantly developing in your backyard. Once these pieces are linked together, we see what we were made to do and how to progress that goal, and we naturally conclude how to live our lives to their fullest.

The name *Quanology*? It's a tongue-in-cheek take on "the Quan" in the film *Jerry Maguire*. The character Rod Tidwell wanted the Quan — his version of this orgasmic life. I thought it was a funny word but the subject is the only thing we care about. It's dead serious and answers some big questions. What is the Quan? Is it the same for everyone? I realized we first have to understand our purpose on this planet before we can answer that. But then, what is our purpose on this planet? We need to care about these sorts of questions because we want the Quan. Even though it may not actually exist, we would all love to have it. Slowly, with each generation, we're moving closer to it. Hopefully, this book will help in that process. Because, knowledge and understanding are great but you need to be able to do something with them. You don't *really* want a map of the world; you want good directions to get from where you are to somewhere better. I'll try my best to provide those directions, but **you** are the vehicle.

Before I discovered Quanology, nothing made much sense to me. I was searching for order and connections in a world that didn't seem to have any. I had investigated various belief systems, but nothing seemed to contain all the answers. And then, I learned about nature. The answers had not only been right under my nose the whole time, they were inside me. But they had gotten muddled in all of society's failed attempts to escape nature. Of course nature holds all the answers to life – your life, my life, and our lives together. How could it not? It *is* everything. But for better or worse, nature – human and otherwise – is not obvious or self-explanatory, and a good deal of clarification is necessary.

In understanding nature, you will come to know yourself better. You're here because you *want* something. You don't want to understand biology – you want to be loved. You want to be respected. You want other people to want you. You want to feel satisfied that everything you do in your life is good and is resulting in something positive for yourself and for others. You want your kids to do well in life, just like you do well in life. You want your parents to be proud of you. These are beautiful things to want, and wanting them has brought you here.

Many years ago, and often enough today, I've needed help. At some point I realized that a good life was not going to just fall in my lap. It was the most shocking and disappointing epiphany. I thought everything would just settle into place, and I'd get all those things I wanted: wealth, family, happiness. All I had to do was hang in there. But no, I was wrong. I had woken up 100 miles from nowhere – down some dark road I didn't even know I was on. How did this happen? I had to do something quickly before it was too late. I had to get back, but back to where? What should I do? The answer turned out to be simple, but I'm what they call a "Doubting Thomas," and I needed proof. I spent years figuring out life just to be able to firmly say, "This is the way forward." The way forward is evolution.

Evolution is the Great River. Quanology will help you journey downstream. Actually, I should say *natural selection is the Great River.* Natural Selection flows like a massive, unrelenting, inescapable river and everything in it will eventually travel in the same direction. It is the force that rules all living things — including us. Natural selection is the process. Life . . . (pause for gravitas) is the result. Natural selection wants all living things to steer downstream. If you steer downstream you will be rewarded. If you try to go in any other direction, eventually you're going to lose.

Natural selection is nature's mechanism for saying that one thing is better than another. The ones that are better stay, and the ones that are worse go away. Evolution is the change that occurs when this happens over time. Applied to life, this means that the animals or plants that do well in their world live, and the ones that do poorly die. Stretched over a long period of time, we can see the evolution of plants and animals – organisms well-suited to live in their environment. As the environment changes, new organisms come and old ones go away. What never goes away is natural selection's judgment – that river never stops flowing.

If scientists argue about evolution it is about details which do not at all change the implication that natural selection created people. Quanology picks up from this point: Natural selection created humans. From this starting point, a flood of insights flows. Through the course of this book I will show you that natural selection still rules over us today.

It's often argued in books and movies and over cups of coffee that the technology of modern life is somehow going against nature, but what you will learn in this book is that technology is natural selection's servant. If we look closely at all the things that make up our modern world, we see that they are all technologies. The story of all life on the planet is a tale of technology. Not just the cars and computers and food processing plants, but the way we talk to each other, the progress we've made in human rights, the fact that we don't kill complete strangers on sight, that we trade, read, cook Thai food, and live better than we ever have before. Technology can be considered synonymous with "good" in this book. It's here to improve our lives and that's the only thing we want.

This book does not promise to deliver your dream life on a silver platter – that self-help cliché. What it will do is change your life for the better by changing the way you see the world and yourself in it and thereby change your motivations and daily decisions. We will take a good look at most aspects of human life: family, politics, religion, sex, murder, drugs, love, money, etc. But you are the focal point.

Why is it always about *you*? The fact is, it isn't. There are other people in your life, other people in the world who you've never met, and your life would not be the same without them. But it starts with you, and really, you are the only thing you can control (sometimes, right?). So, Quanology is a new look at the world, but it's centered on you. It fills the gaps among a number of subjects by providing you with a clear understanding of the world and a clear direction to move toward. I'm going to show you that following your natural evolutionary drives will result in the most fulfilling life possible. "If it feels good, do it" is not as hedonistic and selfish as some have made it out to be. Nature wants us to do things, and it made those things feel good. You can trust her judgment and do them. If we're doing the wrong things, it means we're not feeling properly.

This book is not simple. The Great River is a deep one, so prepare yourself. We're here to get deep because we want to do better. Life is a tremendous gift – beautiful, heart-breaking, and short. We want to get the most out of it. We want the Quan.

A Very Good Place to Start

Let's start at the very beginning. If we are to understand life, understand our purpose and get a clear idea of what we should be doing here, we first have to know *what* we are. **What are we?** This question may seem obvious, but there are many, many different answers floating around in people's minds. It is so fundamental, that you may never have considered that it needed a good answer. We often think more about questions like "What should I do when I grow up?" or "What pants should I wear today?" But that's not starting at the beginning. So, what are we? *We are people* is not accurate enough. We need to know what people are. Again, around the world, you will find many answers to this question. But science has only one.

People are animals. Just like dogs are animals, birds are animals, bunny rabbits are animals, we are animals. Yes, we share many of the same traits as other animals – we move, we eat, we poop, we have arms, we have legs, hearts, livers, lungs, etc. But it doesn't end there. We share the same drives, the same purpose as other animals. No one ever taught me that our actions stem from the same evolutionary motivations as other animals. In fact, they wrongly hinted that humans have an altogether different purpose on this earth.

I thought we were fundamentally different from other animals because we were conscious. But here's a secret: The whole world made sense when I took *consciousness* out of the equation. Pretend it's not there, and you'll start to see things as they really are. See that we act on the same biological drives as all other animals and the sun of understanding will rise for you. We think that because we have conscious thought, we broke off ties with the rest of the animals, but this great myth is responsible for all too much confusion and strife in the world.

Great, we're the fifth ape, now what? This may be an unwelcome thought. I get it, but give it a chance. The top biologists in the world are down with this, so I'm not alone. You may not have heard it from them because they don't really know how to tell you. And really, it's not their job. I'm going to show you that this world makes total sense when you see human life as an unconscious progression in the same vein as animal life. Natural selection still rules over human life, as it always has, despite the fact that we all have cars and internet connections.

When we make the link between nature and humans a lot of things become clear. Roads and buildings are as natural as fields and mountains (they're just not as pretty). Homosexuality is natural. Happiness as a goal will get you in trouble. There's no such thing as peace and harmony. Humans are not completely monogamous by nature. Boys are *mentally* different from girls (physically too, of course). Some people are worth more than others (ouch). Love is selfish, but the "self" isn't what you think. Everybody wants to rule the world, but not just anyone can. War, rape, and murder are all completely natural, but so is love, generosity, and compassion for total strangers. These points still surprise many and make others squirm, but stick with me for another moment.

Let's take things a step further. What do animals want? They want to pass on their genes. Stop. That's all they want. As we have the same motivations as animals, that's all we want. Everything we do is dictated by those little instructions in our bodies called genes. They designed us and everything that's alive. And our genes tell us what to do. We all have stories about why we do the things we do, but our genes know the real reason.

"Stop, stop, stop, John Khoury. I don't want to pass on my genes. I want to graduate from school and get a good job. I want to find a girl and settle down. I want to see the world. I want to have fun. I want to be happy. I want to make a difference. I want to help people. I want to be rich. I want to relax, etc., etc." There is a whole lot of sadness and emptiness in the world because people want things without understanding why. Consciousness does not explain why we get up in the morning and do the things we do.

Again, imagine that consciousness doesn't exist. Imagine you don't really know what you want. It shouldn't be too hard to imagine. I didn't know what I wanted for much of my life. I used to change my story on a regular basis. How about you? Did you think you wanted something for your life, but now you want something different? Do you vacillate on what you want for your life like you do about where you want to go on vacation? If so, you're not alone. If not, you're in the minority. I don't know too many people who have wanted the same thing all of their lives. Most people change, which means they don't really know what they want. Ask someone, "What do you want for your life?" and despite their efforts to give you a nice answer, they will hate you for asking. But here's another little secret: Only nature knows what you want because she makes you want it. And she has stored your true nature in your subconscious.

Like I said, in the past, not much made sense to me. If you're anything like me, there are many theories and belief systems that still don't make sense. This book will explain why. We are going to venture deep into the subconscious and find out what kind of shop nature has set up there. The process will start by understanding what it means to be an animal. Many people are insulted by this very idea. But it is not an insult. It's sheer flattery. Our species has gone from "just one of the apes" to the most powerful form of life that has ever existed (as far as we know). We are *amazing*. Yay, us! If the other animals could fathom what we've done and knew the meaning of applause, they would be clapping and whistling non-stop for us. We are the greatest show on earth! To think that we are made in the image of God, but are just some huge disappointment . . . now, that's an insult. And stop! Before you condemn our species as a murderous, backstabbing, manipulative, planet-polluting, carbon-gas spewing, bank collapsing, donut-eating, lying motley band of selfish pigs. It's just that we do *everything* well – including the nasty stuff. Despite the overachieving, we're still animals, and like them, are bound to the nature that created us and are most fulfilled when we do what nature wants us to do.

Natural selection, the Great River, rewards those that flow in the right direction and punishes those who do not. If you're having a consistently tough life, the reason is simple: you are not flowing in the right direction. Every living thing on the planet has done what nature wanted, the others are gone. Extinct! Fossilized! Eating feels good, sex feels good, love feels good. Friends, accomplishment, raising kids, and giving all feel good, because they all result in the progression of nature's goals. They are flowing in the right direction.

Quanology is about nature – *our* nature and our relationship to the nature that created us. Fulfillment of our purpose is the goal. The reward for going along with nature is a high quality of life. Like all the other animals and plants, we want to pass on our genes. That's all animals and plants want. They don't want to "have fun" or "backpack around Asia," they want to pass on their genes. As you watch elephants on TV or ducks in a pond, you can conclude that all they are doing — the eating, the running, the humping, the pooping, the respective trumpeting and quacking — is all directed at passing on their genes. You probably know this, maybe you learned it in high school, but until now, this truth has never helped you, so you've ignored it. We are no different than other living things in our drives, but we do have more strategies for achieving them. We write operas, we build businesses, we blog, we tweet, we share, we develop, we organize, we protest, we sell cupcakes. In the process we might have fun and crack a smile, but make no mistake, we do it all in order to pass on our genes.

But what does passing on our genes have to do with modern life? We may think of the start of civilization as completely removed from our biological nature, but in the course of this book, I'll show you that everything we did thousands of years ago and everything we do now is still tied to our basic, subconscious motivation to pass on our genes. That doesn't mean we are *actually* passing on our genes, but our motivation remains the same as it was when we were living in trees. How are wars tied to passing on our genes? What about the internet? What about smart phones? What about TV and music? What about sending aid to Haiti?

We'll get to it all. Quanology will provide you with a genetic map of the world — indeed, of your life — with everything on it from your boyfriend to your high school musical to your day shift at the 7-Eleven. Once we have the map in place, we trace a route from here to the Quan. Don't worry that we have not freed ourselves from our natural origins. We never will, but by the end of this book, you will see that we never have to.

The Basics

Before we start this ball rolling, let's all get on the same page here. There are a few basic concepts that the rest of this book will rely on. See if you can handle these. . . .

1. No beliefs allowed

This book uses scientific evidence to understand the world, not beliefs. Oh, we will talk about beliefs and how they work and what they do, but Quanology does not *believe* anything. It strives to *know* things. Little asterisk: We are differentiating between fundamental beliefs, like "God is watching us," and functional beliefs, like "I am sexy." Functional beliefs we get to have fun with and don't need any real basis to them. More importantly, they don't alter our understanding of nature. Maybe you're sexy, maybe not, but great that you believe you are. Fundamental beliefs are quite different. They tell us how the world works and how we work. Unfortunately, they are usually wrong and based on tall tales.

The problem is that everyone has beliefs when it comes to understanding the world which are not so easy to throw out the window. There are usually all kinds of strings attached. Beliefs work a lot like language. As Nicholas Wade in *The Faith Instinct* tells us, just as there is a "language window" when we learn how to talk at a certain age, there is a "beliefs window" when we learn how the world works at a certain age. If you don't learn how to talk as a child, you will never be able to speak well. If you do learn to talk as a child, then the language(s) you learn become a part of you – like an arm or a leg. The same goes for beliefs. If you learn to be a Christian or an atheist or a racist at a young age, the beliefs inherent in those systems become a part of you and you will have a hard time believing something else.

The world is complicated. Theoretically we know more now than we ever have and will know more in the future. But there will always be something in life that we don't yet know or understand. "God did it" has never been a satisfying answer for me – it only added to my confusion. Either we live in an evidence-based world or we can all just make up whatever we want. This book is therefore based on science. Science seeks to understand the world. Nothing more than that. Religion has a different purpose, which we'll get into later. There is no *believing* in science – only knowing and probability. In this book, we'll add up the evidence and come to new conclusions. So, try to put your beliefs aside as you read the rest of this book.

2. There is nothing supernatural anywhere

Supernatural means "above nature." There is nothing beyond nature. Nothing *happens* or *exists* that is not a part of nature. Makes sense to me. It's a pretty clear statement but has huge ramifications. It means that *we* are natural. Just like trees and snails. There is nothing supernatural about us. Even our beloved conscious thought is the natural result of our nerves and synapses. Thoughts and memory need hardware to run – that's brain tissue. If we die, that tissue decays. No brain = no thoughts. It would have been shocking to me to think this when I was a boy. Even today, there are moments when I want consider the idea that there are forces beyond the physical world. But then I remember that it's just my faulty brain on one of its benders. I'm not saying certain things like intuition aren't real; I'm saying they have a natural composition, just like your skin and bones. . . . And you know what this all means for the idea of God? I know, I don't like it either, but that's life.

3. Natural Selection created the world and continues to do so

So, no God. No supernatural forces. What's left? Evolution. Natural selection is the mechanism and evolution is the result it produces over time. It's the controlling force for everything in the living universe. It made us from other mammals. It made mammals from reptiles, reptiles from fish, and so on, all the way back to little strands of RNA in the primordial soup. But we're not talking about your father's definition of evolution, we're exploring how natural selection continues to work its magic in our technologically advanced human society. Human genes are much the same now as they were 50,000 years ago. The difference is in how we *live*. We enjoy a better quality of life than our caveman ancestors, and our lives continue to improve, all thanks to natural selection.

Those are the three principles that will drive the arguments in this book. If I just burst your bubble here, I'm sorry. But I also hope it will result in good things for you. Life is hard, and we are all looking for support because we really can't make it on our own. We are, in fact, not built to make it on our own. The people, the groups, and the ideas that give us support can sometimes feel like the only things keeping us alive. I think you know what I'm referring to. . . .

Religion is the elephant in the room when it comes to the big questions, so let's get it out in the open.

Religion is as big a part of humanity as sex and procreation. There has never been a group of people in all of history who did not have a religion, or at least a belief system which united them and told them how the world worked and how they should live in it. Through studying cultures and evolution and anthropology, I see where religion came from and what purpose it serves, which gives me a bit of objectivity and distance, but ultimately, shining a light on religion, criticizing religion, hurts people. And that's no fun.

I work in technology, so I can imagine someone using a Windows 95 PC to do their computing. It's possible and you'll get most of what you need done. But do I recommend it? Is it not silly to hang on to an outdated technology? When you want to do edgier stuff, you're stuck. If you want to be a part of the latest technologies and information, you will meet with conflicts. It is the same with religion in this day and age. Things move too quickly and technology is too important – the benefits of adapting technology are too great, and for better or worse religion impedes progress in many areas. There is too much *truth* out there – too much information being presented which pulls the rug out from under religions. This information is reaching "the undecided" and they are getting the chance to think for themselves.

Despite its detachment from truth, **religion is a technology**. It seems to be in our nature to use this technology as much as it's in our nature to talk to each other. And just like talking, religion is an outgrowth of our social nature. Couple that with the fact that there are many organizations and serious money behind organized religion, and you can bet it's not going anywhere any time soon. Nevertheless, I'm here to provide an alternative; talking from the perspective of science but treading into realms that religions like to keep for themselves.

There is no God, but there is a force and it's nothing other than the physical forces of the universe setting off a series of events which have so far culminated into a planet covered by humans. We have stepped outside the bounds of biology and crossed into a new plain where our thoughts and ideas are technologies that progress our goal of genetic proliferation. Despite their human origin, it is nature who decides which ones can live and which ones will die. We are being *selected naturally*, just like before, by a force that is greater than ourselves. It is not God, but it created us and we are subservient to it just the same. It is the same evolution that put us in clothes, taught us to farm and to fly and to respect people regardless of their race, sex, or bank balance – all very novel concepts. And it is continuing to take us to incredible places. I believe, eh hem, *the evidence suggests* that with violence declining, poverty on the run, and increasing possibilities to make your life just as you want it, we are heading to a very real Promised Land.

The Star of the Show: Technology

For all its talk about evolution, this is really a book about technology. To transform the old saying "Knowledge is power," let's say that "Technology is power." Knowledge is in your head, but technology is out there affecting the world in your favor. And today you start acquiring a new set of technologies. Technology is integral to human existence, just like hives are integral to the lives of bees. Let me generalize by saying that the person with the best set of technologies and the best drive to use them in the *right direction* is going to have the best experience in life. And technology gets more evolved as we progress. It's (insert current year here) and getting an edge means getting technical. Next year, it will be more so. If you're not technical, don't worry. Quanology will show you how technology is a much broader concept than you ever imagined.

Technology is part of nature. It is everywhere and in everything. It is that which is used to improve. That definition is so fundamental, I'll repeat it. *That which is used to improve.* We'll get into what "improving" is all about, but for now, it's important to understand that technology is not just computers and smart phones. It is a big fluffy tail on a cheetah which helps him to turn at high speeds. It is a cost-cutting measure in the process of making cars. It's knowing to change environments in the heat of an argument to change the dynamic. It is the snap of the wrist when pitching a baseball which makes it drop right in front of the batter. It is applied knowledge of crops which allow us to harvest more in a given season. It's using a confident posture and facial expressions during conversations. These are all technologies. And all around you are infinite opportunities to develop new ones and improve your standing in life.

There is no need to be scared of technology. There is no such thing as a bad technology, though certain technologies do lose their relevance over time. The internet is a technology, but it's also used for kiddie porn. But the underlying technology is still good, it's only the application that can be bad. That's why I said the "right direction." What technology really does is level the playing field. It puts power into the hands of those who have less. Yes, technologies have been created as a means to suppress others – like the nuclear bomb. But the ability of technologies to suppress is undercut as more and more people replicate them for themselves. Look how many countries have the nuclear bomb now. Even "evil" technologies like the nuclear bomb, may have positive side-effects. As author Robert Wright noted, maybe the biggest countries in the world have not fought each other for the past 60 years partly *because* of the nuclear bomb.

Yes, technology is here to help. In order to grab it with both hands, we need a world view which embraces technology and the reality in which it thrives. That world view is based on evolution, and it needs a closer look. Evolution is your mother, and the better you understand her, the better you will understand yourself.

2 | Evolution 2.0

"Darwin's theory is the single greatest idea any human being has conceived."
—Daniel Dennett

We're gonna talk about you. But the most interesting part about you, aside from that haircut, is the fact that you are alive. Since we know that life is not a given, we need to talk about life in the context of the universe. There are many, many planets without life. The moon is not a planet, but us going there is the closest thing we have to visiting another planet. There is nothing alive on the moon. There's not even air or water. It's a just a big rock in a vacuum. Sure, probability suggests that there should be life elsewhere in the universe, but it does seem to be a pretty rare phenomenon. Yet here we are. The forces of the universe made it possible for us to be alive and if you really look at all the factors involved in producing life (which we won't do here), it's pretty astounding to think that life on this planet lasted for more than 10 seconds. In fact, it's been going on for almost 4 billion years. Thanks to evolution, what started as a pond-scum party turned into a planet bursting with an astonishing array of some seriously fascinating animals and plants, all intertwined in a complex web of life and death. But what created evolution?

Big Bang Baby

Quanology assumes that the universe was born with The Big Bang. All matter was concentrated into a very small space and exploded. It has since been expanding – riding on the initial release of that explosion 14 billion years ago. The universe is expanding and everything in it is expanding. This is what *entropy* is – that dissipation of energy from the Big Bang. **Dissipating energy means "spreading things out."** This is the Second Law of Thermodynamics. The inanimate objects you have in front of you, above you, or below you were "put together," but at some point in the future, they will be rubble. Destroyed. It might take 10 years, might take 10 million, but it will get there eventually. What will *not* happen is that those same molecules will remain intact and take on an even more complex form than they currently enjoy.

Take my desk, for example. My desk will not have a nice shine and a new leg 1000 years from now. It will be ruined. That is not to say there will be *no desks* 1000 years from now. There just won't be *this* desk. The molecules which make up this desk will be dispersed all over the world. That's entropy; energy dissipation. **Entropy is happening everywhere all the time. Everything is dissipating energy.** Ultimately, this trend underscores every aspect of life.

Everything goes from a structured state with concentrated energy to a less structured state as the energy dissipates. Try to find a hundred things which could be used instead of a table in this example. Should be no problem. Look around you and focus on something, then imagine what it will look like if it was left untouched for 1000 years.

All the world is moving to a state of dissipated energy. Fill up a balloon with smoke and then put it in one corner of a room and pop it. After the sudden burst, all the smoke molecules in the balloon will gradually and evenly spread out within the room. That's entropy. Pouring a bit of milk into a warm cup of coffee produces the same natural result. If you saw the milk or smoke just hang out where it was set free, you would shout "oh my god" and be torn between staring at it with your mouth open, and getting some friends to show it to. And of course, you know better than to fart in an elevator where the gas's ability to disperse is limited. No, you wait for an open space with a nice gust of wind to conceal your foul deed.

It's a common misunderstanding to see evolution as a system and trend which goes in the opposite direction of entropy. Evolution creates order rather than destruction, right? It takes molecules and makes cells out of them, and organizes those cells into larger organisms. We are seriously ordered. I remember learning anatomy in high school and being continuously amazed at how complex our bodies are. Living things are the opposite of disorder, one might think.

But "disorder" is a bad way to think about entropy, which is perhaps why it has taken so long to make the link between entropy and evolution. Consider a laptop for a moment. . . . This laptop used to be a bunch of unordered stuff, and now, it is a highly ordered machine. As I type, the fan hums, the disk spins, and the battery slowly dies; the laptop is dissipating energy. When we see that the organization inherent in evolution is really just nature finding a better way to dissipate energy, then you see that **evolution, as a biological process, is really just entropy for living things**. There are holes to be filled in the universe, spots where there is low energy, and they need to be filled. Evolution is all about finding ways to do this.

Call it what you will: a hole in the market, an opportunity, a vacuum, an empty spot. These are all temporary situations that will always be corrected by nature. When, as scientists presume, life was sparked billions of years ago when molecules came together to form self-replicating acids, they would only have been allowed to live if they followed the laws of nature. Indeed they are dissipating energy and Entropy thinks that's cool.

Evolution Is the Creator

Evolution is the creator of all living things. Most people understand this. But what still remains a novel concept to most is that it did not create a world thousands of years ago and then hand the reigns over to humans once they could think and talk and make French fries. It is still the driving force behind this modern world. As we said, natural selection is the Great River. It decides what floats downstream and what drowns. If you want to predict the future, you simply need to understand the direction of natural selection and determine what is properly aligned and what isn't.

Being created by evolution implies something absolutely crucial to understanding yourself: **You are an animal whose ultimate goal is to pass on your genes**. More specifically, you are an ape. I know, and you just shaved your legs. This often neglected fact is the basis of everything we experience as humans. Those who deny evolution are therefore incapable of truly understanding human society.

Orangutans are the only apes that live in trees, but doing so does not remove them from the ape family. In this same vein, humans are the only apes that say "waazaaaap" when meeting friends for Monday Night Football, but this also does not remove us from the ape family. Nor do the planes, drugs, credit cards, political parties, or 40-hour work weeks.

When stuffed gorillas were first brought to Western civilization and shown to the masses, people were horrified by the resemblance to themselves. When I watch primate documentaries on the tube, I get an uncomfortable smile when I see how much we are like them. Cut us open and we have all the same workings as an animal. We can replace our hearts with that of a pig and continue to live. We poop, we pee, we fornicate, we give birth, just like those animals at the zoo.

Birth is a big one for us humans. When I was by my wife's side at the hospital while she was in labor with our first born, I remember thinking, "I'm not in Kansas anymore." My wife was no longer a civilized woman; she was an animal. It is in these profoundly "biological" moments that we are confronted by our true nature. My wife went through a metamorphosis in those nine months – like a caterpillar turning into a butterfly. Biology was happening to her. All her education, manners, and grooming could not hide the fact that she was pushing out a new life, just like all those animals. The thing is, biology is happening all the time, we just put a suit on it and gave it a daily routine.

The phrase "monkey in a suit" is actually a huge compliment. We're just animals, yet we've come so far. 100,000 years ago we were nothing special; we were the weird guys in the class that no one knew what to make of. Look at us now. To me, it's more wondrous to see humans as incredibly successful animals rather than lackluster angels. How did we get this far? It's been a long ride. What's important to know is that natural selection did it – all of it. Through slight changes to our biology in each generation, we developed certain capabilities which allowed us to talk to each other and think abstractly. We got hands designed for delicate handling of objects, bigger brains for maintaining complicated social maps, the right set of pipes for complicated speech, etc. etc.

For the majority of our existence, our environment determined how our bodies would develop – just like with all animals. At some point, about 100,000 years ago, a shift occurred and we adapted our environment to fit us. "Enough of this hovering in the rain nonsense—let's build some huts." And "Damn it's cold. I'm going to wear that buffalo." "No more spending days chasing deer in the woods; we'll catch two and make them have babies." Amazing.

But evolution didn't stop. Why not? Because there was still competition. We weren't competing with other species for resources anymore, but we continued to compete with each other. Waiting for biological evolution to adapt our bodies was too slow – a species that can only evolve once every generation is at a disadvantage. Instead we developed technology and the competition then was applied to our technical abilities and collective intellect and could evolve many times in the course of a lifetime. We started to evolve culturally. The game went from "which group or person could win biologically" to "who could win technologically."

This is cultural evolution – the layer above biological evolution. The motivation is the same, to pass on our genes, but the strategy is different. Cultural evolution is the phenomenon of people sharing information and working together to build something which would last for future generations – all these things contribute to a higher rate of genetic proliferation. Killer whales do this; they teach their offspring how to hunt seals in the waves on the shore. The killer whales who don't train their young are at a disadvantage. Chimpanzees do this; they teach their offspring how to use tools to get food. Chimpanzee groups which do not do this are at a disadvantage. But humans teach on a much greater and complex scale. And every year the scale gets bigger and the complexity increases.

Human history is the story of cultures competing with each other, and our cultural evolution began a long time ago when it was dead obvious that our goal was genetic proliferation. Just as the goal of chimpanzees would still be genetic proliferation even if we gave them guns, microwaves, and health care, our goal remains the same. We may have advanced tremendously since the dawn of farming or of language, but the motivation hasn't changed.

Funny how after all these years, evolution is still evolving. The theory is over 150 years old, and we continue to add tiny adaptive chunks to it. Though many people still talk about Darwin, the theory he laid out and the understanding we now have of evolution has changed quite a bit. Everything is subject to evolution – even the theory itself.

In the 1970s, Richard Dawkins popularized the gene-centered view of evolution in his book *The Selfish Gene.*[1] He explained **that it was not individuals who were out to reproduce, but rather their genes**. Genes are the unit of replication for evolution[2]. This added a remarkable twist to the nature of all species. Sterile worker ants could now be explained as carriers of genes which were being reproduced by the fertile ants they

---

1 I know, the last name itself is a slight mutation of *Darwin*. Who will be Dawkins' intellectual successor? Daknil? "*In 2021, Daknil proposed that . . .* "? One can only guess.

2 Nope, epigenetics does not contradict this. Even if genes are being expressed differently in reaction to environmental circumstances, they are still the things that are being passed on or not passed on in the process of evolution.

were supporting. And just when it looked like they were never going to get any respect! We could now also explain how altruistic human behavior could result in the happy death of an individual in support of his genes in another.

If you want to understand human nature, or understand yourself for that matter, you have to become a *gene detective*. We are machines created by genes as a technology for getting passed on. Some genes created fish, some created bacteria, and some created us. In each case, they successfully made copies of themselves and life continued on the planet. Genes all over the world are using their machines as battle gear. Yes, there is a battle going on under our skin, in our minds, and in the connections between us. There is a battle for the precious resources of the earth. The genes need *stuff* to make copies of themselves and they will have to beat out other genes that need that same stuff to make copies of themselves.

You can flick a switch in your head to become aware of this struggle, like you will become aware of your socks as I mention them right now. The genes are constantly fighting, and though we don't see them fighting, we see the results. Like a good detective at a crime scene, we ask what the motive is for the results we see. *How is the gene benefitting?* We have lungs, so we can ask, *what good are lungs?* But take it a step further. We have cars today. Why would a gene want to build cars? How are human genes benefitting from having all these cars zipping all over the place?

As we all know, we have a subconscious brain. We also know that people are often motivated by unconscious drives. We feel a certain way and we don't always know why, but we act anyway. When we are driven to do things emotionally as opposed to rationally, we think it's silly. These "silly" actions are really driven by our genes. Genes want to do one thing: They want to make copies of themselves. And, as far as the genes are concerned, *you* don't really need to know this to go about your daily routine. All other living things have no clue, and they're going about doing it just fine. You are on a "need to know basis." The genes will tell you what they feel you need to know. They want to replicate themselves and you feeling restless, depressed, or horny doesn't require an accurate explanation. Even genes don't consciously *want* to make replications of themselves any more than a drop of water *wants* to fall to the ground. They're just following the laws of nature.

Forget the soul, forget angels and gods and spirits. There's just us. We are animals, surviving by the skin of our teeth. Pull the plug and we become decaying flesh. It may be far in the back of our heads, but we know it. This fact is so basic, but we push it away because, up until now, it hasn't helped us. But Quanology is taking another shot at teaching you about your true nature and finding ways to use that information to create a better life.

A better life starts with a new label for human beings.

You're So Gene-ish

Being an animal who is out to pass on its genes changes the way we see ourselves. We are individuals, but we are also collections of genes. The same genes that are in you and me are in other people too. A gene wants to get passed on and doesn't care who does it. But my genes are also in my family members and my children. **Wanting to do well in life or wanting to feel good are not so much selfish things as they are *gene-ish.*** Remember, the unit of replication, the thing that life itself is based on, is not the self, it's the gene. Nature did not make us to care about ourselves, but to care about our genes. I'll die for my kids, give to my family, etc., etc. I'm gene-ish. So are you, and so is everyone else on the planet.

Being selfish means doing what's best for you. **Being gene-ish means doing what's best for your genes.** You want to live, but your genes have decided you will die within about 100 years. You want to make lots of money, but your genes want you to spend it on other people too. You want to take it easy, but your genes want you to give it your all. Our genes are not counting how many material possessions we have. They are not counting how many kids we have. They are counting how many copies of themselves they are replicating. They don't care about people, they care about activities. "What activities lead to me replicating a lot?" Procreation is a direct method. I'm talking about sex and the babies that come with it. But there are indirect ways like sabotaging rivals, helping out family, and cooperation. These translate into: fighting with and being an average bastard to others, supporting children and other relatives who, obviously, share a good number of your genes, and working together in positive sum activities which result in you and the other party becoming stronger as a result.

This is a new way to look at human beings. We have created a fascinating world through gene-ish drives, not selfish ones. All those inconsistencies you feel when you hear that "people are good" or that "people are evil" all get ironed out when you understand that people are gene-ish.

The Sweet Spot

> "Ladies and Gentlemen of the City Council. I'm just a caveman. I fell in some ice and later got thawed out by your scientists. Your world frightens and confuses me. When I see tall buildings and flashing neon signs sometimes I just want to get away as fast as I can to my place in Martha's Vineyard. I'm more at home hunting the Woolly Mammoth than I am hunting a good interior decorator. And when I see a solar eclipse like the one I went to last year in Hawaii, I think 'oh no, is the moon eating the sun?' I don't know. Because I'm a caveman, that's the way I think."
>
> — Unfrozen Caveman Lawyer

There are a couple of important concepts that I need to explain briefly. You'll see them come back throughout the book. The first is the Sweet Spot.

As I said, we are animals and evolution created us. When a living thing evolves biologically, its DNA changes, which means its body changes. But we pretty much stopped evolving biologically some 50,000 years ago. "Pretty much" because human DNA has changed a teeny, tiny bit since then, but not enough to have a real impact on our lives. This means that a person from 50,000 years ago is "pretty much" identical to us now. We might look different, but the difference comes mainly from the hair care and personal hygiene industries. Look at bushmen or tribesmen who currently don't live in modern society and you don't have to strain yourself to imagine that they could be the same as someone 50,000 years ago. You'll have a harder time with someone on the cover of a fashion magazine.

But, of course, 50,000 years ago is an approximation. It wasn't exactly 50,000 years ago and it wasn't one place. The period of time during which the majority of human life stopped evolving significantly is what I call the Sweet Spot – just like the spot in a tennis racquet's strings that gives you the most juice out of a shot. The Sweet Spot refers to the time when we stopped evolving biologically and started using technology to advance ourselves. Experts on evolution call this The Environment of Evolutionary Adaptedness, or EEA for short, but that is the most unmarketable term I've ever heard, which is why you probably aren't familiar with it.

Life was very different in the Sweet Spot as compared to today. You lived in a group of about 100-200 people, most of whom were your relatives. Men hunted, women gathered, groups fought. There were no pictures, no TVs, no electricity, no phones, no written language, no bikes or cars. Your place in the world did not change in the course of your lifetime, and for the most part neither did your world. Some men in the group impregnated more than one woman, some men impregnated no women, and some men were just dead. By the age of 17, all women were mothers and spoken for. A third of all men could be expected to be murdered, and women lived under the threat of abduction from neighboring tribes. Sounds pretty heavy, but the scenery was stunning!

One of the most important aspects of the Sweet Spot is the fact that men who impregnated multiple women did better genetically. In contrast, women could not improve their genetic standing by mating with multiple men. This situation was a key determinant of our behavior back then. And since we were made for that time, it is still a key determinant of our behavior today.

This was a quick description of a very important moment in history. There will be more detail in the rest of the book, but a full description would obviously be a book in its own right. In his book *The World Until Yesterday,* Jared Diamond talks about the island peoples of Papua-New Guinea who were isolated from the rest of the world until the 1960s. They lived the same way for thousands of years in simple tribes and they were suddenly catapulted into modern life. When we study tribal peoples who have no real technology, we begin to see what we're all about. But this is not easy, since pretty much all people now live in a world based on technology. So, other ways are to study dead societies through archaeology, and to study close relatives like chimps and gorillas or other animals who share similar traits. Not ideal, right, but through multiple sources of information, scientists are piecing together an accurate description of life for early humans.

Despite the persistence of violence and social hierarchy, our world is a very different place than it was back then. Nevertheless, our bodies and minds were made for the life of our primal ancestors. If someone were transported from 50,000 years ago to today, you can bet they would be overwhelmed by everything they saw, just like Unfrozen Caveman Lawyer. We live in cities and pass strange people all day long. We go to grocery stores for food. We can't wait to see what new technologies are coming out next year. We talk to people who live on the other side of the world, and we even occasionally visit them for a week. We climb corporate ladders, accumulate wealth, go to movies, get drunk, watch porn, read, write, build, and use protection, all with the same body and mind we had back in the Sweet Spot. Pretty neat trick, if you ask me. *Bear on a tricycle* pales in comparison.

The Great Loophole

So, I've said that our goal on this planet is to pass on our genes. But there should be a little asterisk next to that sentence. Though the gene-centered view of evolution is now commonly accepted, there are those who try to derail it. Those people will hold up a list of examples of behaviors that do not seem to result in passing on our genes. Let's take a look. . . .

Things we do that do not result in passing on our genes:

- Use contraception
- Study astronomy

Things we could be doing to pass on our genes that we don't do:

- Donate sperm on a regular basis

Actually, we *do* study astronomy to pass on our genes. Women love and readily give it up for men who study astronomy. OK, that's an exaggeration. But we really do study astronomy to pass on our genes. It's a steady job and you might just discover something new. As far as contraception and sperm banks go, these things were not around in the Sweet Spot. And therein lies the Great Loophole.

The Great Loophole means that we were made for a certain place and time, which creates a little disconnect in our present lives. Back then, our goal then was to pass on our genes and all our activities then were aimed at just that. Now we live in a very different place and time, but our bodies and minds are stuck in the Sweet Spot. We are still rewarded by nature for doing things that would have led to genetic proliferation back then. Our caveman ancestors may not have had contraception, sperm banks, or astronomy, but they were trying to advance society by learning things and taking credit for important achievements.

In Chapter 4, we'll examine the Great Loophole in more detail by looking into examples of things we do today that are actually genetic tricks.

But, for now, don't freak out if I say our goal is to pass on our genes. You won't feel fulfilled by selling your sperm on eBay . . . unless you get like a million dollars for it. . . .

Again, we're here to talk about you. We would be fine to keep talking about biological evolution if we wanted to understand you 50,000 years ago, but you live *now*. You are not just biology, you are cultured. Yes, you with your funny hat and multi-colored sneakers. Everything that makes us modern, that helped us get out of the grasslands and into office buildings and McMansions, has to do with technology.

3 | Technology: The Secret Ingredient

Technology is the secret ingredient in evolution. We're going to put a new set of clothes on technology here and show you how all adaptations, both physical and cultural are just technologies. Technologies are anything that improve something. And improving implies a goal.

Who cares about technology? You do. It means everything. It's the difference between us then and us now. It is the difference between people doing well in life and people doing poorly. This book is here for you, and the big theme in *your life* is technology, though you may not realize it. You may have thought it was about *love* or *connection*, but we'll see that those are technologies too. Technology has always been about getting an edge in life – when we learn or create something new, when we find a way to do something better, get a better job, find a better mate, or take better care of our family. We'll even take it back far enough to see that physical adaptations like developing a smaller pelvis millions of years ago is also a technology. In that sense, this game is not new, we just have new strategies and new uniforms so that when we watch Animal Planet and then switch to the news, we don't see the relation.

Every moment nature is evaluating the changing landscape. Are the things you have now better or worse than the things you had two hours ago? What is nature using as a yardstick? In this chapter, we'll have a look at how technology is really the engine behind evolution and what goal evolution has in store for us. Science knows that the yardstick is gene proliferation. Subconsciously everything we do is motivated by this, and technology is what we use to succeed. We can understand this best, as always, through . . . math. Doh!

Evolution & Math

The bad news is that *Life 101* is a math class. The good news is that you can use a calculator. We start with a look at compound interest.

Compound interest

The world is about numbers and compound interest is one of the coolest concepts there is. If I put $1000 in a bank account and get 5% annual interest, my investment will grow to $1050 in one year. We would expect that next year it would be $1100, because we get $50 a year, right? That's what most people who are unfamiliar with the concept think. But really, you get $1102.50! Wow! Don't run off to the shops just yet. I know you think this is not much more, and it isn't, but over a longer time span, the difference between normal interest and this compounded variety gets surprisingly huge. Over the span of 10 years, you'll get $1628.89 instead of $1500. Still not that much, but over 100 years, you'll get $131,501.26 instead of $6000. Pretty amazing, right? I'm not telling you how to get rich, I'm just trying to impress upon you how time and a few small numbers can dramatically change things in ways that run counter to our intuition. We were created to think in terms of months, years, and maybe decades – not millions of years.

Check out this chart showing the difference between simple and compound interest. We're talking about starting with 100 units that increase at a rate of 1% for 1000 years. You can hopefully make out the little line at the bottom that is the simple interest.

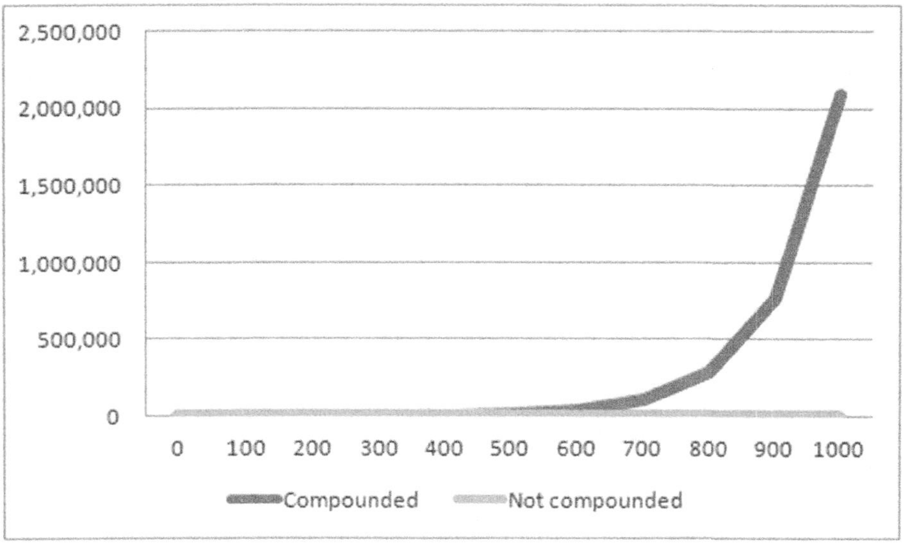

If you can't see it yet, here's a close up of the first 100 years:

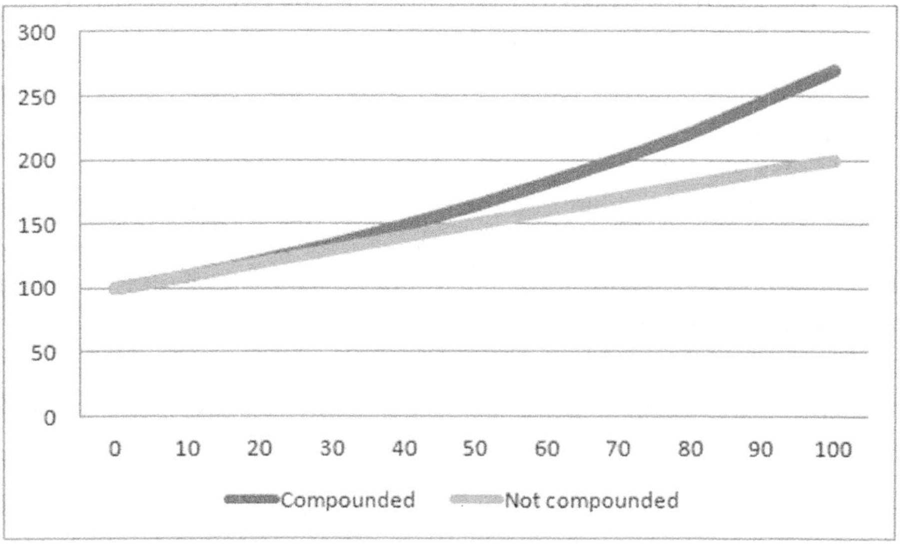

See how much the two lines diverge?

Technology compounds on itself in the same fashion. Robert Wright, in his book *Non-Zero*, notes that life on earth took two billion years to go from eukaryotic to prokaryotic cells[3]. For you laymen, that means going from cells without nuclei to cells with them. Then it took only 700 million years to go from prokaryotes to multi-cellular life. The second transformation took noticeably less time than the first. Each subsequent adaptation takes potentially less time than the one before it.

Why? Because all these adaptations build on each other. That's compound interest. Each advancement in technology makes it easier for new technologies to be built. Just like being able to walk means having more options for food. More nutrition options mean that our offspring will have more viable adaptations. The one with the long neck can be successful. The one who can catch birds can also be successful. The one who can dig well can also be successful. It's like going to a roulette wheel with more chips each time. Before you could walk, you only had one chip. Now that you can walk, the new possibilities open to you are equivalent to, say, three chips. You went from a 1 in 38 chance of winning to a 3 in 38 chance. Yes, competition will see to it that the playing field is leveled again and again, but the point here is that **all adaptations and technologies open up new doors, giving your genes an increased chance of success**. We are now at the point where technologies like the internet and mobile telephones have opened up all kinds of possibilities for us. We can now communicate with people all over the world and learn all that they know in real time. These modern technologies are just as much evolutionary adaptations as the ability to walk. *They are all technologies.*

The acceleration of technology can be compared to dropping a tennis ball from above your head. The first bounce is pretty high, the next one less, the next one even less, until some point the action is moving

---

[3] A life-changing book on how the mathematics of natural selection ensure that cooperation wins out over competition. As the story goes, Bill Clinton made all his staff read the book when he was president. http://www.quanology.org/robert-wrights-non-zero/

pretty fast just above the floor. That is where we are in our evolutionary journey. Things are moving fast today and we have very little time to keep up. In fact, the only thing slowing it down is our inability to process and accept *all* the changes happening around us as quickly as they occur.

Think about yourself for a moment. Everything you have is really compounded on something below it, often from previous generations. Maybe you have a nice job because you worked hard and were able to learn quickly. This may be because you were instilled with good work ethic at a young age, performed well in school and were accepted to a prestigious university. And where did you get that good work ethic from? Your environment. Maybe your parents, maybe a special teacher, maybe something awful happened and it taught you to work hard. These compound experiences led to your current level of success. At each step you compounded, or for corporate hacks, you *leveraged* your position to get a better one. Now, looking to the future, you can continue to do the same. With that good job, you can save money and find ways to help the community. You can attract a good spouse. You can afford to have children and support them in their efforts to get a leg up in life as well. Every resource you develop, every technology you acquire builds on itself. And you're on your way coasting down the Great River.

Investing

"Compound interest and universal expansion? It sounds like the universe is investing?" It is. It expects a return on investment from everything in the form of dissipated energy. It's like a Wall Street fund manager. It has capital, this big ball of energy/mass (remember Einstein's theory of relativity says they're the same), and it expects everything in the universe to invest that energy for it. Anything which does not get a good return on investment will be forced to change form into something that does. If a company is under-performing, then it is forced to restructure or, even worse, go bankrupt; its parts sold off at auction. If an animal is not functioning properly, then it changes form to lunch for a predator, a rotting corpse, or food for scavengers. In every case energy is always dissipated in compliance with the 2nd Law of Thermodynamics. Get it?

We said that entropy is the driving force in the physical world, and that biological evolution is the translation of this in the living world. All organisms also must get a return on investment. Instead of investing money, they invest resources, which have energy implicit in them, and instead of getting money in return, they get success in genetic proliferation.

Example time. Take a male squirrel, for example. The squirrel has a body and a brain. He also has working genitalia. These are resources, and he can invest them in trying to pass on his genes. He finds a mate, seduces her with a couple of nuts, copulates and a few months later, children are born. Now, half of his genes are in a couple of new baby squirrels. In this example, there were resources, they were invested, and there was return on that investment. Let's list the resources:

- body
- brain
- genitals
- nuts
- energy to get the nuts
- energy to find a mate
- energy to copulate

And what has this squirrel been doing up until copulation? Let's take a look at his life up until now. He, of course, started out as a baby, and he had to survive and grow. Some of his brothers and sisters didn't make it. He had to forage every day for food, avoid predators and other hostile males. Food was often scarce – he almost starved. Winters were cold – he almost froze. Over the years he's had a few close calls with predators (like that time your dog almost caught him), and a few beat downs from other men . . . but he made it. Years of struggle and persistence, but he made it. He's tired and hardened, but there's a female and she's not telling him to get lost. Party time.

All this just to pass on your genes!? It doesn't make sense, it sounds like too much work for too little return. If he doesn't want to do all the work, no problem, he can die too. The planet is filled with creatures designed to give as much energy as possible for an acceptable level of return. The animals that give less die out. You were born to give your heart and soul to life. This, ladies and gentlemen, is the great rat race (squirrel race, in this case). These are the things we complain about every day. This treadmill that we cannot get off, which seems to make no sense and have no purpose.

"It's all so much work! To have a life, to have a job, a house, and a wife and kids. It's so much work, and it's not worth it. Screw nature and her plans! We humans want out. We want an upgrade to first class." We've all had a moment when we thought to ourselves, "Life shouldn't be so hard," but there's a danger in assuming that nature is some ideal Garden of Eden and that the mess we're in was created by humans – either stupid, greedy, or both – and if we could just put some honest, qualified people in charge, we'd all be on our way to Easy Street. It's not true. It's a lie sold by politicians, advertisers, and slackers the world over. **Life is hard for every creature on the planet**. That is the nature of evolution. There are no free rides. Yes, some people in life have it easier than others, but not as much easier as you may think. It's often a question of perception.

For me, accepting the world meant accepting that this was the deal. Life is hard. Wake up early, work hard, go to bed late and smile through all of it so that people don't think you are weak. You're going to have to work hard to find a good mate, then work hard to either stay together or break up. And just wait until you have kids. Humans have not reached some exalted plain where we no longer have to justify our existence to nature. Haven't we all thought at some point, "Ah . . . wouldn't it be nice to be a cat."[4]

If biological evolution is not really relevant to people, why are we still doing this? Because now we are evolving culturally. It's no longer

---

[4] Ironically, maybe domestic cats and dogs have reached that exalted plain.

survival of the physically fit, but the culturally fit. Evolution is a constant force, it simply may take on other forms. Evolution says that things compete, and competition is *always* a struggle. You're gonna have to learn to love it, and if you stick around long enough and are open to some of these concepts, you will.

Before you get too discouraged about your fate of hard toil, you need to realize that lots and lots of work has already been done by past generations. Resources have been created for you, technologies developed and handed down. You can use them, but there's a lot to take stock of, and you'll have to study, study, study if you want to make good use of them. There has never been so much potential as there is in today's world. The playing field has never been so level, but there has never been so much to learn to be able to succeed.

Let's take a minute to appreciate all the investments that we and our ancestors have made in the world.

- Language
- Agriculture
- The wheel
- Plumbing
- Sliced bread
- The scientific method
- Double-entry book keeping
- Glass
- Paper
- The printing press
- The combustible engine
- Sheet metal
- Plastic
- Sound and image reproduction
- The light bulb
- The telephone
- Batteries

- Air travel
- X-rays and Ultra-sound
- The internet
- Computers

These things have made humanity robust. The idea that humanity would devolve into a Mad Max world of anarchy and subsistence is all but impossible in reality. All our achievements are part of the collective consciousness of humanity and can never truly be destroyed. Imagine if we were transported back in time 2000 years, it wouldn't take 2000 years to get back to where we are now, but more like 200. The difference is the information we have now that we didn't have then. We invested, and the information technology that came out of it can never be completely taken away. The same goes for you.

Life is not about getting rich, but it is about investing. Every moment of your life, every situation, is a chance to make an investment. Think about all the things you are doing or did today. The obligations are most certainly investments – work, school, kids. If you have something fun planned or you carved out some time to relax, then, hopefully, those are investments, too. The most successful people on the planet spend most of their time investing and getting a good return. Of course, this begs the question: What is a good return?

Genetic Proliferation Rate (GPR) – The New Definition of Success

Let's huddle up for a second. Life is one big investment of energy and success is measured in genetic proliferation. Are we all on the same page here? We need to get this one down before we move on. **Life is about genetic proliferation**. Because of the Great Loophole, *human* life gets to take a vacation here and there, but all life expects genetic proliferation. Life is, after all, biological.

Back to the squirrel. He's been working hard, then gets a chance to mate, scores, and months later his woman gives birth to a bunch of babies. He successfully passed on his genes. Here's a milestone moment where something is missing. What's missing is the equation. Math, remember? Was he really successful? If so, how can we tell? What's the return on investment – in numbers?

If you invest $1000, and a year later you get $1200, your return on investment, a.k.a. ROI, was 20%. Pretty good. If you invest $1000 and a year later it drops in value to $800, then you've lost 20% (-20% ROI). That's bad. The entire global economy, all investors, fund managers, stock brokers, corporate executives, shareholders, private companies, and basically anyone who cares about money, is trying to get a high ROI. We can measure it. At the end of the year, they have a report which has several key numbers:

- Costs
- Revenue
- The difference between these two
- The difference presented as a percentage: ROI.

Every tenth of a percent matters – it's not "around 17%"; it's 17.2%. 0.2% of 100 billion dollars is a good chunk of change, so it's all taken quite seriously. For the financial industry, it is a life versus death equation.

What is the ROI of the squirrel's investment? As far as I know, science does not have a measure for a gene's rate of proliferation. But make no mistake, nature has one. If a gene has a higher rate than other genes, it will exist more, and if one has a low rate, it will exist less, maybe even die off. Unlike nature, we don't have a measure of all genes' ROIs because we can't track individual genes all over the world at every moment.

The good news is that a gene's ROI is always zero or positive as long as it gets passed on by each individual holder. For animals that reproduce sexually, they need to have two offspring, since only half gets into each. That squirrel could work his little tail off and die post-coitus, but as long as his genes were passed in the form of at least two baby squirrels, he gets a positive return. We just have a hard time determining what that return is. We do, however, have some measure of order. If he has one baby in the litter, it's a lower return than if he had 8. If he has poor genes which produced 8 unhealthy babies who die in their first days, he would get a lower score than if he had one uber-squirrel baby who grew up to have many, many uber-squirrels himself. Got it?

Success can be measured in this little game, just not too precisely – despite the fact that the actual score **is** very precise. What's more, the score is always changing. If your kid's genes get passed on, then your score goes up. The difference in your score from year to year is the **Genetic Proliferation Rate**, or the **GPR**, of your genes. Since new life is being born every second and existing lives die off, new genes are being passed on and being put down, so there can never be the necessary freeze needed to check the score accurately. But the point is, there is a score. You have a score. You have one now, and you will have a different one in the future, even past death.

Genetics & Numbers – GPR

Let's say you *could* track all the genes in the living world, count them all up at a certain moment in time and record the numbers. One big Excel sheet/database – column A is gene name or code or whatever (as long as it's unique) and column B is the number – how many there are in existence. Now wait a year and do the same thing again and stick the numbers in column C. The difference between Year 1 and Year 2 is the **Genetic Proliferation Rate** of that gene. It could be positive or negative. The higher the number, the better the gene is doing, the lower the number the worse the gene is doing. Since there are so many genes of any one sort, being that there are so many living organisms, a slight difference in the genetic proliferation rate (or GPR) would have a significant effect on the genes actual numbers. Then do the compounding over a number of hundreds, thousands, millions of years and you get a totally different genetic landscape. And what do you know? The species of living creatures on the planet now are pretty different to the ones millions of years ago!

If this sounds all so technical, it is. I get it. Feel free to re-read it all; I'll wait here.

Genetics can be taken further than you think. It explains everything "living" you see in the world, and that includes humanity. It explains everything you do on a daily basis. It's the reason I like girls; the reason I got married; the reason I had kids and why I want them to take piano lessons and Judo; the reason I cook and work out; the reason I have a job and get dressed up to go to it; the reason I'm writing this book in the first place; the reason I constantly come up with new cool ideas. It's all genetics, and it's all about the numbers. You can deny it in theory, but you can't deny it in practice, lest you – i.e., your genetic makeup – cease to exist. You don't have to "create" kids, but you have to create *something* – it's the only way to improve your GPR and justify your existence to the universe.

The Four Recipes in *The Slight Differences Cookbook*

Genes do everything they can to get passed on and they do this through technology. We can see some of this technology as physical adaptations, but technology also shows up in strategy. A better way of doing something is also a technology. By the nature of compound interest, slight differences between how well one gene is doing as opposed to another can have a big effect. A .0001% difference in ROI on a million bucks over a million years is a huge difference, so every little thing a gene does can help. The genes that used every strategy at their disposal proliferated while those that didn't withered away. And what little things can our genes do? Another snappy title: these are *The Four Recipes*.

- **Sex**: They can recombine with other individuals with good genes and create copies of themselves.
- **Family**: They can recognize themselves in other individuals and spend a bit of surplus energy to help that individual proliferate his genes. Scientists call this strategy *kin selection*.
- **Sabotage**: They can be pricks to other competing genes.
- **Cooperation**: They can work with other genes to get passed on.

The rest of this book will show exactly how these recipes manifest in our lives. Note that there is an inherent preference in these options. Sex is the best thing to do since it most directly results in new copies of your genes. The rest are also effective but they have not directly resulted in your genes being passed on, only in a higher probability that your genes will be passed on.

The Slight Differences Cookbook implies something important about people: they have tendencies. "Humans have a tendency to be adulterous," "Women prefer high status men," "Men prefer young, pretty women," etc. Throughout the rest of the book, we will talk about those tendencies which are universal to all of us. This does not mean we are all adulterous or that all women are disappointed with their non-executive husbands or that all men are disappointed with their less-than-beautiful wives. But trends and tendencies are established by natural selection, and the people who follow these trends leave behind more offspring. But because we are often taught to go against what biological evolution prescribes for us, it is important to flesh out these drives. They are in you, and if you don't recognize them or you deny their existence, then you will never really get a chance to understand society, or to know yourself for that matter. And that would be a damn shame.

A New Look at Technology

Technology is anything that improves a gene's GPR. Earlier I said that technology is "that which is used to improve." But we complain a lot about technology. How often have you heard people say, "*Email gives me less time in the day, not more*"? But "free time" is not the bottom line, GPR is. Now we know that improvement is the same as a positive GPR. Every technology, from couples counseling to iPhones, must be, and already is, graded on GPR just like every financial investment must be, and already is, graded on $$$$.

Technology is the actor in the process of evolution. Evolution is a process during which energy dissipates through technology. Evolution wants the cheetah to eat the gazelle, because doing so would improve the cheetah's GPR. Hence, the cheetah develops all sorts of different physical attributes to become fast enough to catch the gazelle. These attributes — a flexible spine, a fat fluffy tail, big paws, a thin build, a streamlined snout, a big heart — are technologies. *We* are technology. **Nothing exists above the laws of physics which is not some form of technology.** And all of it is helping the universe to dissipate energy – to expand.

Biological Technologies

Understanding ourselves by understanding our biological drives

Like all animals, we are on a quest for high GPR. In that quest, you are the vehicle driving around looking for it. For hundreds of millions of years, humans evolved biologically, just like all the other animals. In recent years, the last two million or so, biological evolution slowly moved to the back seat, and we started evolving culturally. Get it? If the world floods, we are not going to evolve webbed-feet and gills, we are going to build boats and floating cities. Right? Our bodies won't change, our culture will. *Waterworld* was bullshit.[5] We evolve culturally. Currently, this is our preferred method. This book is about you and you will recognize culture in much of your daily life – the talking, the learning, the transportation, the monetary exchange of food, clothes, gas, and gadgets. But to understand ourselves, we first need a good look at those biological drives that are so deep within us. You might be thinking, "I'm not just a bunch of biological drives!" Of course you're not. But they are there and they are strong, for nature spent billions of years building them into you and only a few million creating that cultural shell you now live in. Let us take a little trip down memory lane and see not only how things used to be, but how they still are in those moments when we are purely biological.

---

[5] Look it up. If you haven't seen it, don't. Twenty years after the fact, I still feel like someone owes me two hours of my life. It was an awful movie, starring Kevin Costner, in which the world is covered by water and an increasing number of people start growing gills as a way to deal with their new environment.

All our biological features are there to help us improve our GPR

Remember that biological adaptations are also technologies since they improve our GPR. Multi-cellularity is a technology – billions of years ago all organisms had just one cell. Sexual reproduction is a technology – it allows new-born organisms more genetic resilience. Eyes are a technology – they allow us to critique art.

Of course, every technology has its cost of development. If, at some point, an organism "realizes" that his developments are spending energy and not adding anything to the bottom line, efforts will be stopped. Hippos don't have gills, but they do spend a lot of time under water. Why don't they just develop gills like Kevin Costner?

My children think it is ridiculous that we humans, so advanced and prominent on the planet, cannot run as fast as a cheetah – as if some asshole in Detroit drew us up on a sketch pad without considering what *we* want for our body. But really, our bodies are the result of a cost-effect analysis. We *could* be as fast as a cheetah — any animal could — but the cost would outweigh the reward in terms of that animal's GPR. And GPR is the bottom line. If sets of genes find themselves getting a high GPR, the technologies produced by those genes become a feature of that animal. Apparently, neither gills on hippos, nor cheetah-like speed in boys would improve their respective GPR.

If we are here to talk about our biology and the technologies which are so important to us, we need to start from a basic point and work our way to the more complex parts of our biology. To give a sneak peak, I can tell you that we end up with none other than sex and parenting. Pretty relevant to your life, right? But the beginning is also relevant, though you may not know it. We'll start with parasites.

Parasites

"A whole section on parasites . . . really?" Yes, they're important. The theory, as described by biologist Matt Ridley in *The Red Queen*, is that they are the reason sexual reproduction exists[6]. **We wouldn't have sex if it weren't for parasites.** So, next time you get laid, thank a parasite. . . . Every animal, every plant is a gene replication machine. They were created by the genes to get passed on more effectively. Replicating genes takes energy – just like creating two of anything takes more energy than just hanging out and doing nothing. So, where does this energy come from? Plants use photosynthesis – they take photons of light, mix it with water and CO_2, and take the necessary minerals from the ground and, badda-boom, badda-bing, I have to cut my lawn every two weeks. Animals eat other living things. It goes like this: the plant-eaters obtain the energy that the plants get from photosynthesis by eating the plants themselves. And meat-eaters get their energy from eating the plant-eaters. Got it? Plant-eaters eat plants and meat-eaters eat plant-eaters. Picture the second one. Can you see it in front of you? You're probably seeing an animal being eaten and dying in the process. That poor zebra hee-hawing in pain as the lion, jaws clutched around the zebra's neck, waits for him to suffocate. That's how it works. This is *not* the case for parasites. They eat animals and plants and allow them to live for an extended period of time – all the while sucking energy from them.

As we said, slight differences mean a lot. So, if 5% of your energy goes to some parasite, you are at an evolutionary disadvantage. So what do you do? You evolve ways of avoiding the "parasite payment." Maybe you have genes which have allowed you to get away with only paying 3% to parasites. Well done. Of course, parasites evolve too, so they find ways to avoid your genes which are avoiding them . . . and we're back to paying 5%. And then you evolve again and then they evolve, and this goes on forever. But you are best able to evolve that resistance to parasites when you recombine your DNA with the DNA of someone else – that means sex. In doing so, you create a new "combination to the safe" that the

[6] Check out some info on the book here: http://www.quanology.org/matt-ridleys-the-red-queen/

parasite jewel thieves will have a harder time cracking with their slower asexual reproduction.

This is an example of the evolutionary arms race that Ridley calls "the Red Queen." It is the evolutionary phenomenon where any advantage our genes develop is eventually countered by a competing force. A gazelle gets faster and starts to outrun cheetahs in the neighborhood. This puts evolutionary pressure on cheetahs to get faster. As the slightly faster gazelles start to get away, the faster cheetahs have more of an edge over the slower cheetahs, and before you know it, all cheetahs are now a bit faster than 20 years ago. And it goes in the other direction too. Cheetahs who are slightly faster tend to catch more gazelles, which in turn puts pressure on gazelles to get faster . . . yadda, yadda, yadda. Another 20 years later, all gazelles are slightly faster than before. Red Queen. . . .

This happens everywhere! The Red Queen theory can be applied to the nuclear arms race during the Cold War, for instance. The U.S. had 100 missiles, so the Soviet Union made 200, so we made 300, etc., etc. The same is true of inflation. Unemployment goes down, so now everybody has a job and employers have a hard time finding new personnel, so they have to pay more for new people. That means the cost of production goes up and they have to charge more, which means people have to pay more for stuff. Or, take men and women: men lie about themselves so as to get better women, so women develop better ways to detect lying, so men get better at lying, etc. . . . Red Queen.

It's just another form of competition. And that brings us to the hottest competition of all.

Sexual Competition

Your genes are trying to get higher GPR during sex

So, we have sex as a way to outsmart parasites? How romantic. But that's just the beginning. We are always after a higher GPR. And though competition is a part of it, it's not all of it. I mean, we cooperate too. Competition and cooperation are simply two strategic themes in the quest for high GPR. But when it comes to sex, competition shines very brightly. Why? Because you need a mate, and there are only so many to go around. At the time of this writing, creating mates using a *Weird Science*-esque machine is still under development at Xerox. When it is complete, we can cooperate to develop the production line. Until such time, we will be competing. 3.5 billion boys and 3.5 billion girls, all of varying quality, means sexual competition for the foreseeable future.

But we're getting ahead of ourselves. I want you to understand that this competition is happening not only on the dance floor, but also inside us when we actually make it into bed. Competition, like matter, exists everywhere on all levels. Just like there is big matter and small matter (think planets vs. that small pebble in your shoe), competition exists on high levels and low levels. Nature is constantly judging. That means the living things that want to stay on her good side need to show their best colors and fit well into their environment. But ultimately an environment is a physical space and living things themselves have a physical size. Living things both big and small must compete on all levels, sometimes at the expense of others, for their rightful place on the planet.

On a high level, entire species compete with each other for resources in the world. On a low level, individuals compete with each other for resources. On an even lower level, men compete with each other over women and vice-versa. Have sex with a woman and your sperm will compete with any residual sperm which is already inside her.[7] We're about crotch height now. From there, your sperm compete with each other to see who gets to fertilize the egg. Once the egg is fertilized, genes in the sperm compete with genes in the egg to see who actually gets to be transferred to the new baby. Remember, sperm contains 100% of the father's genes, and the egg has 100% of the mother's genes, so they each compete to get as many genes across as they can, at the detriment to the other person's genes.

---

[7] I know, gross, right? But did you realize that that's why a penis has a bell-end? The ledge on a penis helps to scoop out existing sperm. That seems to be one of the reasons men "pump" so excessively before we ejaculate. We basically spend a few minutes "removing the competition" before we release our own sperm.

Exhausting to read. Exhausting to write. And exhausting to be a part of. But that's competition. That's life. You've played little league soccer, you've bid on some furniture on eBay, you've been "one of four candidates" for a job or a spot in a school. You know that much of your life is competition. If there are activities which are totally unrelated to competition, it is because we have reached a place where we can hang in the mediocre middle. Evolution is still happening to us and we are still being punished for our mediocrity, but too slowly to notice. Some things we clearly and consciously compete for such as money, attention, friends, and the big one: mates.

Falling under biological technologies are sexual ones. Mating is super important. Just ask your genitals. **Without mating, there is no genetic proliferation.**[8] It's step one and therefore, our sex drive is the strongest thing on heaven and earth.[9] Genes in humans, and many other animals, adapt via sexual reproduction. There is good reason why sex feels so good, why we spend so much time thinking about sex, why sex is so taboo and has been so controlled throughout our history. It is the biological a-bomb, the mother of all human activities. Your sex drive is as much a part of your body as your blood and guts. Feeling guilty about having blood? You really shouldn't. Ditto for feeling horny.[10]

But just as survival is not a given for every creature, neither is sex. You have to work for sex. You have to prove your worthiness for someone else to invest their time and energy in you and your genes. Thus, much of our daily actions and the longer term goals they support are really just aimed at sex. We call it "getting lucky," but we really know there

---

[8] For humans and other sexually reproducing organisms.

[9] Yes, it's more the male sex drive which is so overpowering. For women, it's caring for their children. I don't care how big you are: mess with a woman's kids and she will mess you up.

[10] Sexual guilt does have a purpose. There's a difference between feeling guilty about being horny and feeling guilty about having sex.

is much preparation involved. Before you guffaw at this statement, think about how much of your time is spent doing these things:

- trying to lose some weight (men and women)
- worrying about wrinkles (men and women)
- buying nice clothes (men and women)
- trying to look rich (more men)
- trying to look young (more women)
- trying to get ahead at work (more men)
- trying to be good at sports (more men)
- shaving your legs, making up your face (women)
- socializing at places where lots of good looking people hang out (men and women)

It's all done in the name of sexual status. Everything else you do, all those leisure activities, are just breaks from all that "work." Now, sure you think you're doing it for other reasons, but it's just a conscious translation of a subconscious process. Rewind the tape a bit. People have been doing this sort of thing for tens of thousands of years. Maybe they didn't worry about clothes or wrinkles per se, but people have always sought high status positions, and have always tried to look as if they were in their sexual peak and to give off an air of success and wealth. Even back in the Sweet Spot, and they were still grunting back then. You might think, "I want to be thin" or "I want a good job so I can feel good about myself," but those are simple interpretations of the fundamental drive to have sex. Your biology wants you to be/have those things largely for mating purposes.

Your high GPR depends on someone, somewhere with your genes having sex and creating children. Ideally, that someone is you. We are absolutely bound to the first recipe, while the other three are optional. **Someone has to have sex for there to be genetic proliferation!** That's why the drive to compete for mates is so notable and has such a big impact on our society. Sure we competed throughout our evolutionary history for food and social resources, but if you never mate, then these things don't matter so much. This all becomes more interesting when we look at the differences between men and women and how they compete for mates. To understand the *how* we have to first understand the *why*. This means taking a look at the sexual landscape of our lives leading up to the Sweet Spot.

The Human Sexual Landscape in the Sweet Spot

> ***"A man is as faithful as his options."***
> — Chris Rock

What's a sexual landscape? Well, it's not how you trim your pubes, it's the set of conditions that determine *how* we mate. The human sexual landscape is an environment in which competition not only leads to physical adaptations (like the bell end on your penis), but to behavioral adaptions like sexual preferences. All animals have a sexual landscape and a way of mating that is most effective within it. They know what they need to *do*, and what they need to *have* to get a mate. Those preferences for a certain type of mate and the behaviors to get them are hard-wired! They are a part of our bodies like our fingers and toes. **All animals prefer "good mates," but just what constitutes a good mate?**

We were created for the Sweet Spot. Our minds and bodies are now as they were then and the biggest factor determining our behavior back then was the sexual landscape. Thus, the sexual landscape circa 50,000 years ago has the most influence on our behavior and make-up today. Shocking to most. . . and it may take a while for the idea to sink in, but let's have a look.

These are some important characteristics of our sexual landscape 50,000 years ago:

- Men mated with multiple women in parallel
- Women needed support from a man
- Some men were able to support more than one woman
- Babies took many years to be independent
- Both men and women wanted mates with the best genes

What does all this mean? Everyone is out for high GPR and these sorts of characteristics set the boundaries. In each of these items there lies a strategy whose goal is to get high GPR. The first one deserves a bit of a spotlight.

The most significant factor in that sexual landscape was the fact that men could improve their GPR substantially by mating with multiple women. This is a GPR-boosting ploy that women did not and still don't use because sperm are abundant while eggs are scarce. Of course, there are other factors like pair-bonding and parental investment, but this is a big one. What does this mean? Well, first what it doesn't mean is that all men today are working frantically to bed as many chicks as possible (though some definitely are). However, the boost in GPR that it gave has resulted in a *drive* in men to seek multiple women. It's not that men throw out all other strategies in trying to improve GPR, but they have a slight *hunger* for multiple partners that's similar to a craving for a good piece of steak. I don't *need* a steak right now, but I'd love one and therefore might walk that extra block when looking for a place to eat. This drive is stronger in some men, less so in others, and can be affected by environmental cues. But it's undeniable, and it has shaped us. It has shaped us physically by making men slightly bigger and stronger than women. It has shaped us mentally by making men more driven to excel in their careers in order to appeal to women. Mentally men have also been shaped to be more aggressive in order to maintain their position and even fight other men for women.

Boys will be boys

If some men are mating with multiple women, that means that some men are not mating at all. For men, this means a higher reward for being promiscuous and a big genetic punishment for not being so. That competition in the sexual landscape results in the technology of the male sexual drive, male aggression, and the drive to be successful. Like it or not, these things are biologically rooted. If you are the easily outraged type, please realize that they are simply tendencies and neither gender is a black and white picture. But all evolution needs is a mechanism that consistently gives genes a slightly higher GPR, and it eventually builds that mechanism into the fabric of our lives. These mechanisms have been related to sex for the vast majority of our history and are hard-wired into our brains.

What aspects of our society have come about as a result of this sexual landscape?

Check it out. . . . These aspects have been a part of humanity forever. Some you'll recognize; some you'll think no longer apply.

- Polygamy (having multiple wives) is not common, but it tends to spring up whenever possible.
- Adultery by men and women is common.
- There is violence between groups and the spoils of those battles come in the form of women.
- Women need protection and someone to help provide food for her babies.
- Men are prone to kill the babies of other women so as to "free them up" to mate again.
- Both men and women want mates with good genes.
- Men and women pair-bond to better protect their children.

We in the western world have forgotten all about things like polygamy and infanticide. To us, these are ugly actions by ugly individuals who don't fit in society. But I want to change that perspective. It is that our society is so advanced that these things no longer work as a way of proliferating genes. Polygamy is illegal and only people from third-world countries and Utah still do it. Killing your lover's babies is for psychopath males. Killing your own babies is something for the desperate and drug-addicted. But make no mistake, were society to tilt a little in the wrong direction, these things would come right back into the normal course of human life. It is only in the past several thousand years, a mere fraction of our existence, that they have been pushed into the fringe. They are still a part of human nature and are only suppressed when they no longer fit into our modern environment.

You can see some of these aspects as technologies, though we're not used to seeing them as such. Adultery is a technology. Just as a man's body evolves to create millions of sperm to launch into a vagina, his behavior evolves to try to score with more than one woman. A woman will also use this technology as a way to get better genes than the one her current partner is offering. All cases are biological adaptations created through evolution. All are technologies used to gain a higher GPR than other people. Same with polygamy, same with violence. Follow me? Competition leads to technology. Technology results in higher GPR. Such biological technologies are nothing more than instinctive behavior.

"You log out. I'm gonna try to bang your receptionist."
– Jason Sudeikis in *Hall Pass*

Now, I said the biggest factor in the sexual landscape was that men could mate with multiple women. Notice that many of the items in the list above stem from this. **Polygamy, adultery by men, violence, and infanticide are all a part of human life because men can boost their GPR by mating with multiple women.** I said earlier that we are going to be looking at trends, and these are indeed trends. They were trends in the Sweet Spot, and they are trends today. Polygamy has to be outlawed. If it is not, you will get a number of men who take on multiple wives. But you will not find women taking on multiple husbands,[11] and not because they were not taught to do so as children, but because it does not benefit their genes. Adultery is also alive and well in our society.[12] Men

[11] I know, I saw some nutty chick on TV who lived with two men – but this is the point of those kinds of programs, to show you ridiculous things that are mostly scripted fiction. Trust me, the set up won't last long. And yes, those two men are ninnies.

[12] Not an hour ago, I asked my wife, "Who was that guy you were talking to?"

may be just consciously trying to get their rocks off, but their genes are out to create offspring, and the fact that their "replication machine" wears a condom does not change the drive. Impregnating two women is a serious score for the male gene. As per the title of this chapter, it is all about numbers and the GPR is the bottom line when it comes to living things.

Let's do a few junior high math puzzles to be sure. . . .

1. Ricky Sanchez has three babies, what is his GPR (all other factors unaccounted for)?
2. Ben Schwarz has two babies, what will his GPR be?
3. Tyrone Yokogawa has no babies? What is his GPR?

Answers:

1. Ricky has a GPR of 1.5 (or 50%)
2. Ben has a GPR of 1 (or 0%)
3. Tyrone has a GPR of -1 (or -100%)

Surprised? Ben has 0% GPR while Ricky has 50% GPR. The way-too-ambitious young lady in the front raises her hand and protests. *"0% GPR? But they had two kids! 50% in one kid and 50% in another is 100%. Your math is wrong, John Khoury!"* No, Jenny, don't forget he invested 100% of his genes to get those two kids. It's like putting $1000 into a stock, waiting a year, and getting $1000 back. Your ROI is *zeroooo. . . .* If you get back $1500, *then* it's 50%. And that's what the dude with three babies did.

What kind of a drive would nature build into that brain of yours, if a behavior would get you 1.5 GPR? Remember, 1.02 is already a big deal. I'm not saying it's as black and white as "have more than two kids and you're the man," I'm saying it's a touchdown when everyone else is scoring field goals.[13] If, back in the Sweet Spot, you could do it, you *should have done it*, and nature built into men the drive to try.

---

[13] You understand the difference, right? If not: touchdown=6 points, field goal=3 points.

Men also want one good mother for their children

Men are not only driven to have multiple women, they also want good mothers. But we need to understand what evolution thinks a good mother is, not what your mother thinks is a good mother. Since humans pair-bond (i.e. get married), men need to make sure they get the best wife and mother possible. To evolution, a good mother has a few ideal qualities. She is:

- Young
- Pretty
- Caring
- Intelligent
- Faithful

Let's take a look at each.

Young. Being young means she can have a long career of childbearing. Since men were also choosing life-long mates, they wanted to make sure their mate had the most time possible to bear children. So, surprise, surprise, men prefer young women. The dating site OK, Cupid[14] has shared some interesting statistics with the world, which indicate that men of all ages tend to focus more on women between the ages of 20-24. Women, on the other hand, tend to focus on men their own age.

Pretty. What is pretty? It's not a matter of taste as you might think. Pretty, to a man in the Sweet Spot means a few things:

- She has good genes – we see this by the fact that she has a very symmetrical face and body
- She has big hips for easy child bearing (what we call "a nice figure")
- She is young (as we just mentioned), or at least looks young[15]

---

[14] Link to the article: http://www.quanology.org/dating-site-data-mate-preferences/

[15] Being young also implies a few characteristics, such as a nice waist to chest ratio, big eyes, nice hair, etc. Even being blond is associated with being young. This is why men like these characteristics and we call them *pretty*, because they are traditionally more common in young women.

Caring. This means she will be a devoted mother. Women, of course, have different degrees of parental involvement, just like fathers. You know that some women around you are more caring mothers and some are more "will you shut up, I'm watching *The Price is Right*!" mothers. The first version is more attractive to men.

Intelligent. Yes, not only can smart women handle themselves well in life, they have smart babies, so men are attracted to women with brains.

Faithful. As we'll soon get into, an unfaithful woman brings a potentially huge threat to a man. This being the case, his psyche has a clear distinction between women he thinks will mess around and those who will be faithful.

In the Sweet Spot, men who mated with women with these traits got a higher GPR than those who didn't. This is why men like the women they do. Women, in turn, try to present these traits when looking for men. Today, men might not be looking for mothers for their children, so it really makes no sense to be interested in 16 year old girls since you would be crazy to want to get one pregnant. But you should understand by now that this logic means nothing to a brain that was built for 50,000 years ago. This is what the sexual landscape demanded that men want from women. **Not only have men developed a preference for these sorts of women, they have an evolutionary drive to compete for them.** Because how many women do you know that are all these things that will talk to you for five minutes?

What about women?

> ***"Fuck me, Your Majesty!"***
> – Annette Bening in *American Beauty*

Likewise, the sexual landscape has built behaviors and preferences into women. Women, like men, want good mates. Women shop around too, and seeing as they don't get higher GPR by mating with multiple men, the one man they do get had better be of the highest quality. **Women are choosy**. We men know this, and we suck in our guts when they walk by. But there is a difference between landing a man for sex and landing a man for life. The man you most want to have sex with may not be the man who is best suited to help raise your children. And the man who is most willing to help you raise your children is not likely to be the best man genetically to father those children. After all, the men with the best genes are more capable of landing multiple women, so why would they choose the strategy of settling down and committing? They wouldn't be doing their GPR any favors.

Women want the best of both worlds

This little chess match has provided women with a sneaky move of their own: they can settle with a dependable man of slightly less genetic quality and find the genetic quality they need in hotel romps with their tennis teacher. Again, this is about trends and tendencies under the surface. Are all women pursuing this strategy? No, but maybe 2% are. And what did we say about 2%? It is a big number to evolution, big enough to shape society.

Sure, you're a good girl and would never consider such a thing. But if you're a lady, close your eyes and imagine you could have an affair with someone. What kind of guy would it be with? Someone with good or average looks? Someone with a powerful or an average job? Be honest. It's just you and me here, so no need to be shy. I know, it's just a fantasy and you're not going to do it, but maybe there's a little part of you that would like to, and now you know where it comes from: your evolutionary drives. It's not because you're feeling fat or your dad didn't hug you enough. These drives are on a much lower level. We can make laws about adultery, the major religions forbid it, Muslims cover their women in burkas and take away their right to drive a car, but what remains unfettered is the drive that evolution gave to us, and no law can undo it.

Women are caregivers

Those helpless babies need care, and women are the sex chosen to give the most care because they are the ones who have invested the most in having the baby. This means that women are nurturing by nature. Again, it is not something society has taught women to be, it is something evolution has *created* women to be as a technology for getting higher GPR. Men provide more care for babies today than they ever have, but the behavior in both sexes was not built for today, it was built for 50,000 years ago when it was quite clear that women were taking care of the kids and men were off hunting.

And that's what we are all about: the drives of the Sweet Spot. Your brain was built for then, not now. **Understanding you *now* means understanding you *then*.** And so goes the story line of evolutionary psychology. It is the *why* of psychology, and it explains human behavior quite well.

Despite rationalizations for why men are so ambitious in today's world, we can conclude that they are still doing what they do for those same two reasons. **Men are ambitious because they want higher quality mates and better care for their children**. And the same is true for women, despite what they'll tell you about wanting a guy who is sweet and funny. **Women really like guys with good genes and high social status**. If they meet such a guy, they will surely rationalize that he is also sweet and funny, despite the general consensus that he is neither. Women compete with other women for those high-quality mates. And, like men, women strive for good care for their children. This all seems to be pretty universal. If you don't like these ideas, I understand. I'm the sweet type, so I don't like it either. How dare women say Handsome McSquare-jaw is just as funny as me, 'cause he's fucking not. Just admit that he's boring but you want to sleep with him anyway. I would love some scientific evidence to show women that their preference for Ryan Gosling and multi-millionaires is the result of brainwashing. But it's actually quite natural. And I'm sure there are plenty of women who are happy to tell their man that their monogamous relationship should keep them satisfied for the rest of their lives. Under pressure, these men will agree with them. But the things we consciously want and the things that are biologically natural do not always square up.

Women have a drive to seek both good genes and a good provider.

Today, women are more than able to raise a baby on their own. They can even go to a sperm bank to find the best genes possible for their baby. But back in the Sweet Spot, it was a different story. They needed the support of a man, and as those drives carry over into today, women still largely seek men as long term partners and feel incomplete without them. Back then, and even today, those men provided food and protection. Without that support, women in the Sweet Spot were at a horrible disadvantage so nature built into them a great fear of being left alone by their man.

While the drive for good genes makes women attracted to good looking guys, the fear of abandonment drives them to seek the dependable types as well. The best women will have better options for both, and a lucky few will find both in the same man.

But once the babies are born, she can put down the "good genes objective" and focus more on the care. Many women do — they cut their hair and put on a few pounds as they are no longer pressed to look as good as before. They'll make good use of that freed up energy to take care of their kids and support their community.

We know mothers as the caring, loving people on the planet. Men, too, can and will be loving parents, but they never put down the quest for good genes entirely, so forgive them if they are a little distracted at times and maintain their drive. Nevertheless, our society finds a solid backbone in the care the entire family unit performs, and this, too, is part of the sexual landscape.

Family Matters

There are more ways to get your genes passed on. So far, we've focused on mating. But there are more recipes in the cookbook. The sex was fun but, of course, after every party there is "cleaning up." That means child rearing. This is the *Family* recipe.

***Family* means to care for those with the same genes you have.** Genes want to make copies of themselves, but like a good team player, they don't care if they score the goal or if one of their copies does it – as long as it gets done. Genes are built to help out individuals who share their genes. Your siblings and your children have 50% of your genes. The only kinds of people who have more than this are identical twins (100%). Since your children theoretically need your help more than your siblings, you are highly motivated to care for your children. This has been carved into your subconscious over and over again for tens of millions of years.

Let's face it, you don't *have* to raise kids. You could just have them and hope they take care of themselves. But evolution has shown this to be a poor strategy for humans (turtles are great at it). If we leave them to fend for themselves, they will surely die, and all that work you put into getting a mate and bearing a child will have gone to waste. So, **we are a species of highly-involved parents**. This has also been carved into our subconscious . . . perhaps not as persistently as finding a good mate, but enough to shape our world significantly.

Genes want to make copies of themselves, and if they don't produce direct offspring, they turn to the strategy of caring for copies of themselves in others; namely family and even friends.[16] It's called kin selection, and the concept goes back as far as Darwin himself. Of course, he didn't know about genes back then but the gene-centric view of evolution has only given this theory a more solid foundation. That's the second recipe in our little cookbook.

***Sabotage* is the third recipe.** For men, sabotage comes primarily in the form of violence. For women it shows up in the chatter, gossip, and

---

[16] Yes, it turns out we even choose people who have similar genes as friends. http://www.npr.org/blogs/health/2014/07/14/331354227/do-we-choose-our-friends-because-they-share-our-genes

teasing; these are ways to cut your competition down to size without having to use violence. Violence is mostly a guy thing, and we've already alluded to a number of sources of male violence. Though people in the Sweet Spot lived in fairly egalitarian societies, our evolutionary history has not always been egalitarian. Men had to fight for high status in their community. In our tribal way of life, violence existed both within an individual group and toward other groups. But the more energy devoted to internal competition, the less you have left over for external competition. And our move toward that egalitarian life in the Sweet Spot meant transferring our aggression from internal to external settings; we fought less within the group and more with other groups. This was a good move, but it was still violent, and it still required men to be strong and aggressive. So, let's see, war with other groups, the need to be able to defend your own children from the infanticidal drives of other men, the need to ward off other men from your woman, etc. **All these things have led to aggressive men**. Implicit in all these aspects of our evolution are ways to sabotage other genes.

Women also adopted the strategy of sabotage. Whereas men are motivated by multiple mates, women are motivated by quality mates. Since quality mates are in short supply, the women who did their best to get them had a higher GPR than those who didn't. **Sabotage is a perfectly valid strategy for women, too.** For instance, I'm sure you've observed young, unmarried women occasionally being bitches to one another.

While men compete physically, often outside the group, women do so emotionally and socially within the group. Rumors and gossip about a girl who did this and that with so and so spread quickly and affect that girl's mating potential within the group. **It is not just a cliché that men prefer virgins as long-term mates**. A virgin has no chance of tricking a man into raising someone else's baby. A woman who has slept around has a *slight* chance of doing just that. Now take that word "slight" and give it a big hug, because it is so important to evolution and the development of our behaviors. We're taking from *The Slight Differences Cookbook* and it all revolves around consistent little boosts or drops to our GPR. So a woman who has slept around a bit, even if it's just a rumor, will have a *slightly* harder time finding a committed man. And she really wants that commitment. The women who could best manipulate social impressions and statuses were best able to sabotage their competition. So it was in the Sweet Spot. And today? Are girls and young women spreading rumors about each other? Again, not you, right? You don't talk trash about other girls, but maybe you've wanted to. Maybe you've said things like "not putting her down, just saying" or "I feel sorry for her that she needs that confirmation from other guys" or "I wouldn't be surprised if she did sleep with him."

Like our sex drive, sabotage is a biological technology which improves GPR. We are hard wired to be jerks under certain circumstances. The men without this technology did not do as well in competing for women all those years ago and thereby died out, so over time, evolution made more men violent. But as violence became focused on external groups, internally we were able to spend our energy on other things.

And that brings us to **the fourth recipe: *Cooperation***. Working together demands different things from our brains; it demands a cooperation drive. After all that fighting and gossip, you'll be happy to know that people also have a biological tendency to work together. Groups in the Sweet Spot were egalitarian because we *could* work together. The apes that don't cooperate much (all apes other than humans, that is) have a stricter hierarchy. The guys and girls at the top are doing most of the mating. But with cooperation, the guys at the bottom can gain strength in numbers and also strength in the technologies that result from cooperation. That means they can have better mates too. Cooperating means being friendly and understanding the wants of others. It's a nice addition to all of our personalities, and one we'll soon discuss in more detail.

Looking at the sexual landscape shows us that being a good mate is important and being a good parent is important. But *life* in the Sweet Spot shows us that it's also important to contribute to society. Being good at these things means being good at passing on your genes. Can we measure *good*? I think so. And in doing so, we're going to get some real insights into human nature and our own subconscious actions.

For so long, the world overwhelmed me. There was so much I didn't understand. Why were people treating me the way they were, and why was I treating them the way I was? Why did it look like some people were getting a free ride while others were denied the things they wanted? I know the self-help industry will tell you that the world treats you the way you tell them to treat you. But that is too vague for me. Some people are good looking, some are smart, some are rich, and some are strong. All these shapes and no puzzle to place them in. Through Quanology I saw that it all fits together.

Every day I try to grow, to progress my life. I know you do too – though some days are better than others. Now I understand what *progressing life* is: It's getting high GPR. Nature makes me feel *complete* when I do this. Competition means that some people are doing well and some people are not. But we're all so different, so how do you measure who is really doing well? Nature has an answer. Yes, it's another number.

GPP: The Value of a Human Being

My father used to put his thumb to the front of his ear and his index finger just to the top of his head and tell me, "A man is measured from here to here," alluding to intelligence as the measure of a person. Is that right? What about looks? What about money? In my experience, most people have their own ideas.

I remember the moment as a little boy when I realized that humans not only have value but have *different* values. Someone on the news was in a life-or-death situation and I was amazed to see that the world did not stop entirely to save him. At that age, that's what I expected to happen, but apparently the normal functioning of society was more important. I wonder if it had been someone else, maybe someone famous, would more of an effort been made to save the guy? We'll never know for sure, but we can have an idea.

Investing is about increasing value – acquiring the energy that evolution wants you to acquire so that you can disperse it, as entropy demands. In living things, that means getting high GPR, which is to say, to pass on our genes and raise our offspring to do the same. **Our capacity to increase our GPR is a measurable value.** Of course, we can't measure it like a bathroom tile, but we can define the traits that compose it and give them an order of value. We call this order ***Genetic Proliferation Potential***, or **GPP** for short. The GPP measures how well a creature can pass on its genes.

Imagine it's 10,000 years ago, and we don't have a yard stick yet, so we can't measure distance accurately, but we know the difference between far and close, long and short. From where you are to the moon is far. From where you are to a tree is probably not too far. In the same spirit, we can estimate someone's GPP by degrees. As I was thinking up this concept, I thought of Paul McCartney. He is the equivalent of "here to the moon." He is the pinnacle of GPP. It won't take a long argument to figure out why: He has an outstanding reputation, based on the high opinion of him shared by hundreds of millions of people (many in very high places), several high-quality offspring, good looks (don't forget he was the cute Beatle), serious wealth, personal charm, a positive attitude, and is the best friggin' bass player of all time. He has so much going for him, most of you didn't even know about that last one. What about more common people? A pretty young girl? A wealthy man? A smart person? Fact is, we can compare. All other things being equal, who has the higher GPP in the following examples? I know you don't like judging people – you're a good person. But just think objectively. In terms of evolution, who among the following people would be better at passing on their genes?

A person in the city or a person in the country?

A woman in her 20s or a woman in her 50s?

A man who is 6'3" or a man who is 5'5"?

A person with an Ivy League education or a person with a high school diploma?

A person in good shape with a nice physique or someone who is out of shape and overweight?

A person with a high IQ or a low IQ?

A man with Kennedy hair or my hair?

A person with good social skills or a person with poor social skills?

Are all men created equal? How about all women? We know the answer to this. We would like to say "yes," but we know the answer is "no." Do we really not want to judge people? We do it all the time without thinking, but if you ask us, we will say that it is wrong. Or maybe we just don't want people to judge *us*.

We are here to understand, not have things told as we would like to hear them. You might be disappointed, but when you understand that that's the way things are, you can come to peace with them. If we suspend our judgment about judging people, we know that a Kennedy baby is not equal to a baby from the Balderschmidt family in Dundalk, Maryland.[17] We all know this. It is hard-wired in our brains to judge people on their "level."

Here's the good news. **Genetically, we are all created *almost* equal.** The slight differences concept means that we are all genetically only slightly better off or worse off than anyone else. . . . Little Todd Balderschmidt is not as genetically far off from baby John F. Kennedy as our intuition might suggest. If his genes were so inferior, nature would have eliminated them long ago. Obviously, his genes have an almost equal chance of getting passed on as JFK's. Thinking again back to the Sweet Spot, a person's job was to get a high GPR. And their ability to do so can be projected in their GPP. This is the new measure of a human being. We've had "height and hair," we've had IQ, we've had EQ, now we have GPP – Genetic Proliferation Potential. You don't have to write it down; just highlight it here in the book. Definitely share it on Facebook and Twitter. . . . I'll wait. . . .

---

[17] I sincerely apologize to any and all Balderschmidts in Dundalk or anywhere else. I tried to come up with some fictitious name, but got unlucky.

People are created almost equal in terms of GPP. All animals species on this planet share the same inequality. There must be some inequality, since if we were all genetically equal, there would be no evolution – nothing to select in favor of something else. **Evolution exists only in a universe that is slightly *unequal*.** As the universe expands, it changes and that is the space in which evolution works. Who or what can best fill the gaps caused by that change?

So, *genetically* all people are created almost equal according to this new measure of a human being. But then, where do our differences come from? Everything else. Everyone is born with a certain potential genetically, which is slightly less than equal. But the set of circumstances into which you are born is **not *at all* equal** and, in fact, can be really crappy or really great. And as circumstances such as how well off your family is and what area you live in change over the course of your life, so does your GPP. Now we start to see the real difference between Todd Balderschmidt and John Kennedy. Todd is from a lower middle class family in a poor suburb of Baltimore. The schools are not that great in his neighborhood, and his peers are not very well educated or wealthy. There is a lot of riff-raff and even violence. What's more, they have no real access to people who are educated, wealthy, or connected. Contrast that with the Kennedy's. John was born into a wealthy, well-connected family. The town he grew up in had some of the best schools in the nation. He was close to Harvard, one of the most valued higher education institutions in the world. Having so little violence in his life, he was able to focus on developing himself. The difference in GPP between these two individuals is obvious.

Here's an exercise. I call it the "hanging off the side of a building test." Imagine both men are hanging from the side of a skyscraper, about to fall off at any second. You can only save one. Who do you save? If you're honest . . . we're going to miss Todd . . . he was a good man . . . salt of the earth . . . his death so tragic. It's ugly, but it's honest. As creatures born to spread our genes, we know who to ally with and who to stay away from. I'm not saying it's the way it should be; it's the way it is.

I know this could be construed as cynical, but if you feel guilty, try to suspend judgment. It's all going somewhere good. In fact, we are always sizing up one another, and until GPP, I never understood the yardstick we were using. Why were some people seemingly getting special treatment – getting saved from falling off a building, while others were left for dead?

Not only are you born with a set of circumstances that influence GPP, but things in your life can change and affect it, even in an instant. Imagine two unheard of tennis players in the final at Wimbledon. The morning of the match they may be equal in the world's regard for them but that night, after the match, it will be a different story. What happened? Nothing changed in them. But one won the match and the other lost. In the morning they had the same GPP, and after the match one of them had bolstered his GPP. The other did not.

"I'd be struggling to find a girlfriend"
– singer/songwriter Jake Bugg, when asked what he'd be doing if he weren't a musician.

GPP is only so complicated in humans. Modern culture and technology have provided more options and opportunities to proliferate our genes, and they all have to be accounted for. GPP in most animals is purely biological and predetermined by what they are born with. Not so in humans. But the goal remains the same – high Genetic Proliferation Potential (GPP) means a high Genetic Proliferation Rate (GPR) – at least it did in the Sweet Spot. Don't forget that GPR is the proliferation rate of an individual gene, but GPP is *your* potential for helping your genes get along. You may have 700 brake horsepower under the hood, but doesn't mean you are driving 200 mph. You are the vessel and your genes depend on you to drive it effectively. They are working together in the form of "you" to get passed along into future generations. That's all any animal or plant is – a coalition of cells who hold the same genes working together to get those genes passed along.

We are animals, but unlike them, we can evolve more than once in our lifetime. This is an often overlooked distinction between humans and the rest of the living world. A fulfilling life will depend on doing just that – finding ways to use modern, cultural technology to improve our lives. "Do what nature says and use technology to do it well." It will be the people who best employ technology — not the smartest, richest, or the best looking ones — who have the best experience.

I think back to an interview I once saw with former James Bond actor Pierce Brosnan. He was telling how being so damn good looking was a detriment to his acting skills. Being good looking meant he got parts when he wasn't actually the best actor. He was not forced as hard as other actors to hone his skills, and he lamented this. Traditionally, it is not the chiefs or strongmen of the tribes who develop the technology and take our society to new places. It is not the pretty women who charm us with their personality or who make the best mothers. It is the ones who most need technology who are most prone to using it. Before we get into just how technology has an effect on our resources, let's take a better look at those things that we all have that define our value.

Resources

What are you made of? You've got arms, legs, eyes, ears, and a nose. For the most part, all people have the same make-up: bodies and brains. Some nicer bodies and better brains, but we all have them. Our genes created these bodies and brains as a better way to make copies of themselves. Bodies and brains are themselves technologies, but since we all have them, we refer to them here as *resources*. There are other things we all have. We all have friends, we all have families, we have a place we live, we have things we own – all of these to varying degrees. But all these things help us to do well in life. They are the resources we use to succeed.

Your GPP, your value, is made up of your resources and the technologies you use to put them into action. Think of resources and technology like a car. All cars pretty much have the same parts – chassis, wheels, motor, brakes, seats, steering wheel, transmission, gas and brake pedal, etc. It's the technology that can come out of nowhere – like turbochargers, fuel injectors, ABS, and power-steering which can add greatly to the performance, comfort, safety, and fuel efficiency.

Let's suppose you are a car and your goal in life is to be the best car in the world. Your parts are pretty much the same as every other car. If they were too different, you wouldn't be a car any more but rather a boat or a plane or a skateboard. Right? Cars, like cougars or cheetahs, are all pretty much made up of the same parts.

What are *our* resources – our parts — as people? Basically every human being has these resources in common, which contribute to their GPP:

- Sex organs
- Physical appeal
- Social network
- Energy and physical ability
- Location
- Assets
- Knowledge
- Intelligence
- Thoughts

Important stuff. We all have these resources, and their combined, possibly weighted, value plays a big part in our success in life. Technologies will be the tools you use to either improve one of these resources or to better put them to work in your life.

Let's take a detailed look at each resource.

Sex Organs

We start with your fun parts. As I said, at the foundation of GPR is having babies and you need working equipment to do this. Not everyone wants to have kids, and that's fine, but let's not look too far ahead. We'll get into the meaning of all this in a bit. For now, as far as your genes are concerned, your working genitalia are the most important resource you have. When they take a hit, you feel pain. When they have fun, you have fun.

Physical Appeal

The Brad Pitts of the world, the Scarlett Johanssons—they are considered the top in terms of physical appeal. Beauty may be skin deep, but it also seems to be universal. Despite the fact that you and your friends may argue over who is better looking, generally "attractiveness" is not a matter of taste. People are either "hot or not" and then it becomes a question of personal taste, often subconsciously directed by the genes themselves, when you decide which of the attractive candidates you would choose to be with.

When studying beauty in all animals, scientists boil it down to symmetry. All physical appeal really is is a genetic doctor's report. It is showing potential mates the quality of your genes. You are supposed to be symmetrical. If you are not, then your genes were not able to deal with possible parasitic infection during the building of You. This asymmetry serves as a warning to potential mates about issues their offspring might have were they to copulate with you. "You're too good for him, girl." Birds do it, bees do it—all animals are judging others on their physical appeal, implicitly asking the question, "If we had kids together, how strong and fit would they be?" Because we all want the fittest kids possible.

And then there is the youth factor in attractiveness, as we just talked about. There is an age range when people are the most attractive, and we can translate this into being the "most virile." Men can be virile quite late in age which is why you hear women say they don't mind a touch of gray on a man. Attractive women are usually younger. Men's genes want lots of kids, so they are more attracted to younger women because younger women have more time to have many kids.

Social Network

It's not who you are, and it's not just who you know – it's who you know and what they think of you. A monumental concept is that anything and everything you want in this life, someone else has to give to you. A good job, a family, money, fame, sex—someone else has to give it to you. You'll have to work for it, but success is dependent on what you get from others. People are takers, but they are also givers by nature; it's part of their life strategy. They give most easily to family, and next to people they personally know. And then, to people they have met in social settings. If they like you, they will give to you. But we need to understand what it means to "like" someone. In this context it means that they are worthy of giving to.

Go check out *The 48 Laws of Power* from Robert Greene and you will notice that those laws are all about how to deal with other people. It is a book on social strategy. Those other people hold the key to power in your life. Your social network is a big deal and much of our behavior is geared to improving it. All the little things we do and many of the big things are efforts to get on the mental map of the rest of the world.

Physical Fitness

As I haul two suitcases out of a train on my way to the airport with my wife and kids, 30 pound backpack on my back, I am reminded that life is a young man's game. I'm in good shape. I don't think I'm a happier person because of it, but then I don't know how I would be if I was fat and out of shape. But I'm differentiating here between looking good physically and being physically fit. We may think members of the opposite sex are attractive, and our opinion of them counts as a resource for them, but the ease with which they use their bodies to accomplish life's physical tasks is a different resource altogether.

And life requires physical tasks. It's a contact sport. Someone with lots of energy and physical strength is going to get through life easier than someone who has a hard time getting out of a chair. Other people, the people who we said are the ones who give you stuff, know this, and they value physical strength and will consider you more valuable if you have it. I'm not saying you have to be a body builder, and this does not apply as much to rock stars or painters, but your physical fitness will be noticed and it will play a role.

Location

"Thanks to our parents for raising us in New York City."
– Beastie Boy Ad-Rock, in his Rock and
Roll Hall of Fame acceptance speech.

Remember USA for Africa? People starving in Ethiopia? There are good places to be in this world and there are bad places to be. I'm not dissing Ethiopia, it's just that they were a bad place to be for a while. At the time of this writing, there are a few towns in Syria that have come to a standstill. Work, school, transportation, you name it, has been put on hold because of violence and civil unrest. Within a country, there are good and bad places to be, and they can depend on what it is you're trying to do. Easy example is wanting to be a successful actor. Successful actors have great GPP! Why wouldn't you want to be one, right? Well, if your location is Allentown, PA, you're going to have a hard time. You need to be in LA or New York. Allentown is closer to New York than say, Little Rock, AK, so in that sense, the location is not a disaster. But high energy, big cities attract high energy, ambitious people. Again, depending on what you want to do, living in such a city improves your GPP.

Assets

"Money can make a bald spot into a 'part.'"
— Chris Rock

Everyone has assets and/or liabilities. Do you own stuff? Do you have money? Do you owe people money? Your financial status will have an effect on your GPP. Women are attracted to money, and men are therefore driven to get it. Women want money to help care for their children. Same for men. Women can also have assets and be rich, but traditionally it has been men who have sought money, largely because women expect that of them. It has been shown that even rich women like men with money. Men do not place much value on the wealth of a woman though obviously a rich woman has a higher GPP than a poor woman, all other things being equal.

Knowledge

> ***"Knowledge is power."***
> – Superman

There is certainly a good bit of truth to the statement "Knowledge is power." If I know something that you don't, doesn't that give me the upper hand somewhere? Of course, that knowledge has to be relevant and it has to be applicable somewhere. Knowledge does fluctuate throughout your life, and its value too. Some things are good to know at certain ages and other things at other ages. Knowing how to catch a football is good to know as a teenager but less important as an adult. Tax laws are good to know as an adult but not so much as a teenager. We forget things as we go, and we learn all kinds of unimportant things which only clutter our brains, but in general, someone who knows a lot fares better in the world than someone who knows little. Ignorance may be bliss, but again, we are assuming here that nature doesn't want you to meditate with a smile on your face all day long.

Intelligence

I hesitate to put this in its own section, seeing as we already have knowledge and thoughts are up next. All intelligence does is help you get knowledge or develop thoughts. But it is not knowledge itself or thoughts themselves. Therefore, who cares about intelligence if it's not resulting in knowledge or constructive thoughts? Still, intelligence consistently shows up as a valuable resource to people. It adds a degree of flexibility to knowledge and thought. Creativity stems from intelligence and being creative helps to gain new insights when solving problems. You can see intelligence as a resource which helps you get other resources.

Notable when taking intelligence into consideration here is the conclusion by many scientists that intelligence is only valuable up to a certain point. Apparently someone with an IQ of 120 is just as capable of being successful at a task or career as someone with an IQ of 160. You reach a point where "more" doesn't matter. Back to cars – a car with 600 bhp is not necessarily slower than a car with 800 bhp. It might be were they to race in a straight line, but then life is not a straight line, and other factors mean just as much as intelligence after a certain point.

Thoughts

Some people have better thoughts than others. That is, the thoughts they have give them better results in life. This is the world of Cognitive Psychology. Two people in exactly the same situation with the same resources we've gone through can have two very different experiences. This ultimately boils down to the thoughts running around in their heads. Yes, it's "looking on the bright side," interpreting an unclear situation as to illicit positive results. There's always a downside and a bright side to things. The relatively new field of Positive Psychology is studying those with positive thoughts and concluding that they lead more fulfilling lives. Of course, their definition of a fulfilling life comes at a different angle than the one we are assuming here. For now, the person who thinks "I can do anything" is better off than the person who thinks "I'm no good."

Why do we care about resources? Because our genes want to be passed on and having a good score in these areas leads to this. To translate to modern times, if you have a high quality or quantity of these resources, you will have an easier time in life. We all want to be good looking, well-connected, wealthy, physically fit, smart and positive. Why? Not because these things make us happy, but because they help us get what we want. We want to be around other people who are good looking, well-connected, wealthy, physically fit, smart and positive. We want them to like us, and more importantly, do things for us. We don't care what those *other* people think of us.

While you were reading this, I will assume that one or more of these resources made you feel uncomfortable. You are not strong in that area, but that's OK. In practice, most of us are **not** good looking, well-connected, wealthy, physically fit, smart and positive. Are we screwed? Thankfully, no. Just as a car has a set of resources, you have a set of resources. And just as a car can employ new technologies to improve the performance of its resources, so can you.

Cars 100 years ago still had wheels, seats, a motor, a windshield, and brakes. Cars then and cars now both want to drive, but the difference in how they drive is dramatic. Today, the difference between people with a high GPP and a low GPP is largely technological. *Are you infertile?* Go to a clinic or adopt a child. *Not the best looking guy/girl?* Put on these pretty clothes. *Not so great with the opposite sex?* You can employ better communication techniques. *Poor?* Work with a team of people to build a business. We all have the same set of resources, but we can improve those resources through technology.

Technology – Your "Better Half"

Your genes built you with all these resources to give you a better chance of passing them on. Slight differences create lasting advantages, so any aspect of who you are that gives you an edge will be passed on to the next generation.

500,000 years ago, a person's resources were pretty much stable from adulthood on. You're 14 and whatever you are, whatever your place, that was it. All your resources were pretty much static. With the onset of culture, that all changed. Humans reached a moment where the things they learned could be passed on to others. Now not only were the best genes being passed on from one generation to the next, the best tools and thoughts were being passed on from person to person within the same generation.

The evolution of our complex brain developed in lock step with the ability to communicate those thoughts through complex language. In this development, we entered a new race. The goal was still the same: high GPR, but now people could do something new that added a whole new level of strategy. Yes, biological adaptations can be seen as technology, but in humans it works on a whole other level. That level is cultural. Technology is *our* way of improving our GPP. All those resources we just went through can be improved through technology. The end goal that nature wants from us still remains the same: to improve the proliferation rate of our genes. Technology is your friend. **Understanding people means understanding technology.**

Technology & Culture

Technology is the secret ingredient of human civilization. It has taken us from a roving band of hunter/gatherers on the African savannah to a bunch of assholes on yachts.

How did we get technology? Silly question, you might think, but do we really know? Can we imagine technology as something as natural as a bee hive? Or imagine a road or bridge or smartphone as natural as anything biological evolution ever created? I've always felt the split in people's minds, and certainly my own, that life in the animal and plant world was somehow different from life in our world. Nature was on one side and everything that we created was on the other side. But I've since learned that technology is welded to nature. For humans, nature is one side of the coin and technology is the other side. So let us again look at technology, not as something different from nature, but as an extension of it.

I also hope to show you that technology is the way for you to become the best version of you possible. **Your fulfilled life will depend on**

technology. Our lives have always depended on technology and it is in our DNA to use it and enjoy it. Yes, your computer is technology, but so is meditation! So is yoga, so is mindfulness, and so is exercise and good eating habits! Seeing these things as technology will make you conscious of their power as tools in your shed to develop the best *you* you can be. What is the difference between a highly-evolved, fulfilled human being and someone we call *savage*? You'll soon see that the real difference is technology, and the result is a difference in the value of that human being. A person loaded with technology has a better ability to proliferate his/her genes while a person with little technology has a lesser ability to do the same.

In the eyes of nature, technology is nothing other than an adaptation. We can imagine elements of an animal or plant as a sort of technology: the ability to move toward light in plants is a technology. Imagine a plant that could do this while the others couldn't. Its genes would have an advantage, it would spread more, and soon enough, all plants could move toward light. No problems so far, right? The technology here is natural, is it not? Good. So, we take things a step further.

A male gorilla lives in a group under the tyrannical rule of a large silverback. The silverback hoards all the females, so none of the other males father any of the children in the group. The silverback has seven children, the other males have zero. Therefore, the silverback has a very high GPR compared to the other gorillas. Any attempts to flirt with the females results in a thorough beating by the silverback. One day, a subordinate gorilla sees a sharp stick, so he picks it up, and the next time there is a scuffle, he jabs it into the silverback's neck, twists it, and pulls it back out. Technology. Still with me? This is a *cultural* adaptation. Why cultural? Because, the knowledge of how to use weapons is not passed on through the genes. If this stick-wielding gorilla has children, which he now may very well do, they will not be born knowing how to stab someone in the neck. This adaptation, if passed on at all, will be passed on through culture – people who know stuff teaching those who do not.

"See this stick, son? This stick is the reason you are here today. Let me show you what you can do with this."

Can we see that this too is a natural technology? It's a friggin' gorilla. It doesn't get much more natural than that. Same as with the plant, a technology was used, it resulted in improved GPR for the individual that was using it, and natural selection gives it a seal of approval. Plants move toward sunlight, and gorillas use weapons.

Of course, gorillas don't really use weapons, but humans do. So, in that last bit substitute "early humans" for "gorillas." And now we start to see the beginnings of our technology as an evolutionary adaptation – all with the same goal of high GPR.

Each type of technology is in service to our genes. We – as groups and individuals – gain an advantage over our competition from our abilities to use these technologies. This all started a long time ago but ever since we have been building towers of cultural resources. By employing cultural technologies, we went from cave men to Yuppies in red pants and set up shop all over this planet.

Technology was able to structurally increase our success rate as a species because we are social animals. Remember, slight differences. A 1% increase in GPR is a big deal. What do the genes want? High GPR. And 1% is a big fat roast lamb to a hungry gene. Any technology we create that serves up a slight increase in GPR is going to stay. Living in groups gave us many opportunities to create them.

Living in groups means communicating and sharing with other members of the group. It's not a given – not all animals are as social as we are. Hell, not all apes are as social as we are. Just look at orangutans – at parties, they don't talk to anyone. If we didn't live in groups, there would be no cultural technologies. But the group lifestyle also comes at a price, since it needs to be maintained. Groups have to share at least some common goals and they have to work together to get them. Otherwise, what's the point of being in a group? If there were no common goals and no improved result in getting them, nature would favor solitary life for humans.

The quest for higher GPR through information technology is cultural evolution. All of culture is about bits of information. Anything that is passed on from one brain to another counts – and all bits of information are judged by natural selection just as all genes are judged by natural selection. They are only allowed to exist if they help their holder pass his genes on (or at least not preclude him from doing so).

Get Smart

You have to talk to be able to share anything beyond basic information. The better you talk, the more complex information you can share. So, of course, in the beginning, speech was crude, but it developed. People could start to learn from each other in more meaningful ways. Not only "how did you kill that buffalo?" but "show me how you made that knot." People were already working together at this point, but now through communication, we could take the good ideas that different people had and spread them around. The inventor of an idea could give it to someone else who could improve it. "Here's a little stick I use to catch fish." "Why not make it curved so it goes up through the fish's mouth when he bites it?"

Man-made things are physical but what's not physical is knowing how to make them. Once you know how to make them, they can be duplicated. That's where the real value comes from: the instructions. It doesn't cost much effort to exchange, reinterpret, and improve information, but the benefits of doing so are huge.

Information has sex too

Just as our genetic information is shuffled during sexual reproduction to create something slightly different from its parents, ideas are slightly recombined when people discuss and transmit them to create slightly different ideas. The ones that come out worse are killed off by natural selection, and the ones that come out better improve the fitness of those who hold those ideas. Really, this is all innovation is. Matt Ridley describes this phenomenon as bits of information "having sex with each other."

You may be thinking that comparing sex and innovation is a stretch. But they are both technologies that improve your GPR and they both feel good on some natural level when you do them right.

Technology in the form of information means that you have an evolutionary advantage over someone else. It means your group has an evolutionary advantage over another group. The value of a person or group used to be stored solely in their genes, but now it is also stored in that technological arsenal we refer to as culture. 50,000 years ago nature deemed all our genes pretty much equal. Now we see differences between individuals in the form of their technologies. Even that new resource, wealth, must be managed or else it will disappear. How many people come into money, have no idea what to do with it and then lose it? Many. . . . My dad once mentioned on old saying "an ounce of money requires a pound of brain." **It is still the technology that you need to get ahead – for individuals and for groups.** Gaining knowledge is a key aspect to fulfillment.

We understand that *know-how* has an intrinsic value. You are better off for knowing things, but you only know as much as you do because you live in a group. Information gets passed on and passed down through generations. Every new generation does not start at zero knowledge. 99% of the stuff we know, we learned from others. Groups aren't the answer for every species – leopards don't live in groups and they are doing just fine. The benefits of a group must outweigh the costs. What are those costs? Getting people to work together means they have to do a few things. They have to be able to communicate. That costs energy. They have to be able to split up tasks, as in "I'll vacuum; you put away the clothes." That costs brain power. Then they have to be able to split up the rewards of their cooperation, as in "one for you, two for me." And, of course, they have to agree to do the same things. This is a huge problem! My wife and I often argue on Sundays because we can't agree on what to do! Clean the house, go for a drive, have friends over? What about bigger groups? Is all the fighting in America the result of a big group of people who can't agree on what to do? Can't agree on who does what? And can't agree on how to split the benefits of their efforts?

Surprise, it's always been this way. These problems are inherent in groups and are a big drain on GPR. So, nature had to come up with a way to keep groups running smoothly. The more everyone is on the same page, the higher the GPR will be. The inherent fighting within groups represents a gaping wound that bleeds GPR, so nature invented the mother of all Band-Aids: Beliefs.

Beliefs

***"I'm sorry, Elaine. I cannot be with someone who doesn't like** The English Patient."*

– Elaine's boyfriend dumping her in an episode of *Seinfeld*

As we crossed over from the rule of biological evolution to rule by cultural evolution, we changed. We became connected to each other in a much deeper way. Beliefs became the glue of that connection. We could work together to create things that we never could have created alone. We could fight together against other groups more successfully because, even if we didn't share the exact genes, we shared the same belief system.

The whole point of culture, of the group nature of humans is that there is strength in numbers. This strength comes in two flavors. Firstly, when fighting, it helps to have more guys on your side. Secondly, being in groups offers humans opportunities for win-win situations –"Non-Zero Sum games," as Robert Wright calls them. Yes, competition is everywhere – all the time in all creatures, but so is cooperation. And before you think it's some warm-fuzzy aspect of nature, you need to realize that it is simply another strategy your genes use to get pushed into future generations. Genes can encourage their human vessels to cooperate with other humans and get a slightly higher GPR than if they did not do such encouraging. Sure, the other human would then also benefit, giving his genes a slightly higher GPR, but if both ended up with higher GPRs than humans who did not cooperate, then the behavior stays. It becomes an adaptation, and voila, we are cooperators.

Beliefs in small tribes originally kept us from getting hurt and kept us together as a group. So, **we believed that we needed to fear things that might hurt us and we believed that we were better than members of other groups**. These are two universal properties of beliefs. And, again, when I say beliefs, I mean thoughts that are suppositions of the truth. The fact that they are beliefs means they cannot be proven. It's quite possible that God will strike you down if you utter his name. Maybe he just hasn't done it the past 1000 times, but the next time he will. And it's quite possible that your favorite sports team is superior to another despite the fact that they were beaten this season. Next season will surely show their true colors. The point is, beliefs like these helped early man to be careful and stay together. And, of course, beliefs like these are still present today, not because we are just as stupid as back then, but because our brains were made for back then, and so they keep using those same mental paths.

Beliefs also motivate. Through motivating beliefs, we get out of bed and work for our destiny. In the face of regular fighting with other groups, the groups who were most motivated and tightly knit had the advantage. They fought more fiercely than other groups. In the chapter on rituals in *The Faith Instinct*, Nicholas Wade points out the difference between two groups of people. Group A regularly performs ceremonial rituals which lead to strong bonds. Group B does not. In the throes of battle, Group A will keep fighting with only 10% of its soldiers still standing. Group B will retreat once 10% of the group has been killed. What a staggering difference that bond makes. As my buddy Rich used to say, "It's not the size of the dog in the fight; it's the size of the fight in the dog." Beliefs are that *fight* inside of us.

Groups need glue. Every company, every army, every soccer team needs a hierarchy, needs rules, processes, and common goals. Without these things, we're just a bunch of people bumping into each other. We often don't like groups because of compromises we are forced to make to be in them. Often those compromises come in the form of beliefs. "Just because the group believes it, doesn't mean I have to." But now we see that this is part of the deal. Beliefs are group technologies that provide a better way to be ***one***.

Even without the fighting, you need to be one. As we all know, there will be enough struggle within a group for a better spot in the hierarchy. Cooperation may be everywhere all the time, but so is competition! It's better for a group that dissonance be kept to a minimum while aggression is kept facing outward at other groups. When groups get too big, the genetic relationship waters down and people need something extra to keep everyone on the same page. And that's where religion comes in like a hammer to a board full of protruding nails.

For the most part, early civilizations were composed of tribes – groups of 100-200 people who were all pretty much related on one level or another. So, altruism was natural. "We take care of our own" and such. Kin selection – that theory of genetics which explains that we more easily offer help to members of our family, implies that people in such groups looked out for each other. If all these people in a group are related to me, then my genes are in all of them, and I am encouraged by those genes to help out fellow members of the group. The better they do, the better my genes do – even at a detriment to myself as an individual.

Of course, such a group is limited in its power by the number of individuals in it. Like a corporation that wants to get bigger, groups in recent millennia considered mergers. You find other tribes and you want to gain the advantages of positive transactions with them. The problem is, there was a natural tendency to want to war with them too. I mean, you could engage in positive transactions by trading, but you could just as well kill them and take their stuff.

Religion, in this sense, became a strategy. Robert Wright points out that two groups who did not share the same genes, could now share the same beliefs. Beliefs then acted as a sort of "gene substitute" – a shared replicable unit of commonality that each individual held which would allow them to at least be cordial to one another. "*Let us not fight, we're both Jews,*" for example. The stronger the shared beliefs, the more they trust each other and the more effective the group will be. Beliefs, packaged up in religion, were developed to facilitate this.

Groups needed to grow if they were to survive and sharing beliefs allowed them to do just that. Those who didn't grow, who didn't use beliefs to merge with others were taken over or killed. The winners were now poised to innovate and prosper on a level never seen before. But remember the real winners are not people and not groups, but genes. The people who were happily led to believing things had a huge boost in GPR, and those who did not easily believe things saw their numbers drop. Soon their owners would be big enough to form cities and societies and take a gigantic evolutionary leap away from the hunter/gatherer lives of those before them. All thanks to beliefs. It was a question of life and death and it is only via the technology of beliefs that it is possible.

Some beliefs are good, some beliefs are better

As we mentioned before, **beliefs are only adopted if they contribute to higher GPR**. All "living" religions believe that it is your duty to have many children. Some dead ones may have believed otherwise, but that may be the reason they are dead. The Catholic Church is against contraception because it hurts their numbers. That's the only reason. What the church leaders wanted was "*frequent copulation of married Christian couples.*" Somewhere along the line the idea that "*sex is a gift and God intends for you to use it for procreation*" was attached to the desired behavior and you get the result that Catholics are against contraception. But we Quanologists know the real reason they hate contraception – it's not a sin against God, it's a sin against their organization and its longevity. And rightfully so, I might add. Nicholas Wade speculates that this is exactly how Christianity came to be the dominant religion of Europe – they simply *outgrew* their Jewish and Muslim competitors. Remember Ricky Sanchez had three babies. You need *more* than two babies to grow your gene pool, so any group of people who thought two was the perfect number of children died out.

Along the same lines, the Protestant work ethic is great for GPR. I once saw a program about German farmers (yeah, back when I had free time). They were working hard in the fields on a horrible winter day. It looked so grueling but one of them solemnly explained, "A day of hard work is like a gift from God." That's much better for your GPR than "Screw this, let's sit inside and get drunk."

Beliefs are always doing several things:

- They are motivating people toward productive behavior
- They are bonding individuals into groups
- They are telling us how the world works
- They tell us we are *better*

Beliefs are also judged by natural selection

Unbeknownst to the prehistoric groups of many thousands of years ago, some beliefs, as with any technology, were better than others. They weren't consciously selecting their beliefs, they were just making stuff up to the best of their ability. But natural selection was judging those beliefs. I can't say specifically what all those beliefs were at the time – they preceded the written record — but imagine a really bad belief, like say, *having more than one child makes God angry*. So people would kill any second children they had. But this was before medicine, so many children would have died before reaching sexual maturity and eventually the population of the group would fizzle out to a small number of people, who would then be ripe for take-over by another group. The group dies out, their genes are replaced. And the belief? It would go extinct with its holders.

What's a better belief? "*We are God's divine creation and he wants us to have many children and take over other groups*." So the members of the group all go to it and each woman has, on average, 10 children. The group grows dramatically in number. They have more people than they can really handle, so they're up for putting the young males on the battlefield. All these males feel in their bones that God is cheering them on. If they die, the numbers are culled to a stable level. If they win, they take over another group and reap the spoils of victory: cool stuff and more women to have more kids. The group does well, the genes do well, and the belief? It spreads like a weed.

We see that cultural evolution is the process of ideas, beliefs, thoughts, and information developing over time through natural selection. It is not us who decides if a thought, idea, or a belief is worthy of existence, it is nature.

Today, the same game plays out in the beliefs of the world. The original tribal groups have grown into countries, religious organizations, political parties, and activist societies. They all hold beliefs that tie their people together, and their beliefs are now involved in competition. Unlike other forms of information technology, beliefs don't have to be true, and by definition, they really aren't. Similar to other forms of information technology they just have to improve the GPR of the group. As far removed as we think they are from our animal roots, the drive is still genetic proliferation. If gorillas formed a political party, we would know what their end game would be: improved GPR for its members. Despite our progress and conscious thought, humans follow the same incentives.

Forming groups happens automatically

And now the population has taken sides and people are grouped. It's "us against them." It's not a physical competition, it's a cultural one. In all different levels, people are grouping up and forming subconscious coalitions. At work and at school and in the neighborhood. But the winners get the same spoils as in a physical war, only to a slightly lesser degree. But that's OK, no one had to die in the struggle – just die off. Imagine if the whole country was all Republican or all Democrat. You might think that that would be the end of political struggle, but no. . . . Eventually, we would gaze in judgment at our neighbor countries and engage the war of beliefs with them, again with the same goal as our recent triumph over the political opposition. And of course, this is what dictatorships spend their energies on – looking outward. Democracies have internal struggles, so they think less about taking over other countries. The moment everyone within the country is on the same page, they will start spending their energies on expanding outside the country. Struggle never goes away. It's the nature of beliefs to compete with each other, just as it is the nature of genes to do the same. Ultimately, beliefs and genes are the same thing in different forms, and both are judged by natural selection on how well they fit into their environment.

Belief's Success Factors

What makes a belief succeed? I can think of three factors:

- How easy it is to understand
- How easy it is to pass along
- How well it sticks with the consumer (like, how well it affects emotion, or how personal it feels, or how entertaining it is)

Sounds like a virus, right? A successful virus needs easy entry to a host, must be passed on easily, and must be able to stick around long enough to be passed on.

A belief like "*Jesus loves you*" is a world-class belief. Even if you have never heard of Jesus, it's easy to understand. It's easy to pass along. And it grabs you right by your heart. Love, like a parent's love, means total support. And aren't we all, regardless of age or status, dying to take a break from it all and have our parents "make it all better"? "Jesus loves you. He'll take care of you no matter what." Sticks like gum on a hot sidewalk.

That last factor is ultimately about emotion. How well does it engage the emotion of its consumer? Emotion is the glue of learning. If something affects you emotionally, you will learn it more easily. It's like your brain prioritizes information and the more emotion associated with a piece of information, the higher the priority it gets in your head.

Beliefs are not true

The fact that "the truth" is only one of many factors in the success of a belief is a strange concept at first. By definition beliefs are not true. So, why do we bother with them? Why can't the *truth* unite us and do all those other things beliefs do? Because the truth is complicated. It cannot easily be explained in a mouthful of words, so one must create sound bites which give the impression of being true. "*We are here to love one another*" is not really true, but it passes easily from one mouth to another's brain.

At the point when we were able to think abstractly and communicate, we also realized there were some things we just didn't know, and that scared us. Most things actually. So, we postulated. We hypothesized. We guessed. We proposed. Out of those postulations and guesses came beliefs. Assumptions. They probably weren't true, but that didn't matter. We now had beliefs and these beliefs not only provided us peace of mind, they banded us together. "*Where did the world come from?*" "*The God, Zul, created it.*" "*Why does the wind blow?*" "*Because Zul is trying to cool off a pizza that just got out of the oven.*" "*Why are we here?*" "*To have children and rule over the earth.*" If a group all believed the same thing, they were able to get on the same page and work together better as a group.

If truth is not the only factor involved in accepting information, then the door is open for the "things we want to hear." We complain about politicians who lie, but we the "thought consumers" have subconsciously demanded this, since it's the lies that pass muster on the three factors, not the truth. The truth is complex, too long-winded to pass along, and appeals mostly to our intellect, not our emotion. The truth has to fight with its hands behind its back!

Take presidential elections, for example. You're sure to hear both candidates say that "this election is not just about who is president for the next four years, it's about the future of America." Some TV pundit will say "If Hilary wins, it's the end of the Republican Party." These statements don't have to be true, and of course, they aren't, they are simply rocking the three factors.

So, what about the truth? Does accuracy of a belief hold any advantage? Yes and no. Accuracy is not a factor in the success of a belief, only the three factors listed above. But the second factor has a hidden track – you cannot pass on beliefs if you are dead. Therefore a belief is also successful if it contributes to the longevity of its consumers. This means that accuracy can hold an advantage since a belief that holds a better perception of reality can help its holders deal with reality better. Notice I say "can" rather than "have to." The benefits of accuracy need to be looked at on a case-by-case basis. Believing your group is "God's chosen people" is inaccurate, but it doesn't seem to be a disadvantage to believe it. On the other hand believing "snakes are to be handled as a show of dominance" is inaccurate, and holding this belief is a disadvantage in the long run. This is why the snake-handler's religion doesn't see its numbers grow.

You'll notice many groups switch seamlessly between valuing beliefs and valuing truth. One day they belief that God created the world and throw science out the window, the next day they get sick and run to the doctor. You can call them hypocrites (and Lord knows I have), but they are just doing what is natural. In this way, they get the best of both worlds, and their GPR is boosted by both ends. On the one end, beliefs are working their magic, on the other end, information is improving knowledge and leading to technical abilities.

But there is another strategy in using information.

Your Thoughts in Other People's Minds

If genes are units of information in a biological world, then thoughts are units of information in a cultural world. We know that genes want to create copies of themselves. What came as a surprise to me, is that thoughts want to do the same thing. Like the genes inside you, the thoughts in your head want to get out into the world and multiply. I've known about this for years, but it still strikes me as new and fascinating – even as I write it now. **This is the drive we all have to *be someone*, to share, and to influence.**

In an episode of *Curb Your Enthusiasm*, Larry David spent a minute complaining about how Christians want you to be Christian too[18]. He says, "I like lobster. Do I go around pushing lobster on people? Do I say, '*You must like lobster*'? '*Eat lobster! It's good, it's good!*'" But in reality, this is natural – as natural as wanting to have sex. We want others to believe what we believe, to like what we like, and to think what we think. We may deny this consciously, but remember the gap between what we consciously and subconsciously know. In fact, Larry David *does* want you to like lobster. If you go out to dinner with him, order first, and order the lobster; he will like you more. We all have thoughts, ideas, beliefs, and knowledge – these things amount to our cultural make-up, like our genes are our biological make-up.

Renowned biologist, Richard Dawkins, called these units of information *memes*. Throughout the rest of the book, I'll use memes interchangeably with "thoughts," "beliefs," "ideas" and "information."

Think of memes as little workers created by genes. Imagine there is a father who works for money, but one day decides his ten children should go out and do the working for him. They go out into the world and give him 20 percent of everything they make. Some tell him to bugger off and give nothing back, but in the end, instead of the 100% he was making by working himself, he gets 20% x 7 of what his children are making – 140%. Pretty good deal.

This is essentially what genes have done in creating thoughts and beliefs. They are bits of information still tied to their creator who go out into the world and get support for their creator. At some point they can cut off ties and get separated from their masters, like when no one remembers where a thought or good idea came from, but this was not the master's intention when he created them. If I ever have any good ideas, I want people to know they were mine. Don't you? Go watch the "Anonymous" episode of *Curb Your Enthusiasm*. Ted Danson still finds a

---

[18] What a funny show that was: http://www.quanology.org/larry-david-comparing-lobster-to-christianity

way to take credit for anonymously donating to a project, and even gets the added prestige of having done it anonymously. Like "what a selfless guy." Speaking of *Curb Your Enthusiasm* , its creator, Larry David, in an interview said how rewarding it was to finally get recognition for his show, which was something he felt he missed in the *Seinfeld* days, when he worked behind-the-scenes. In the long-run, anonymity stinks.

In today's social media landscape, why do so many people spend so much time and effort posting their status to their friends and colleagues? In light of all the above, you should be able to see the real reason behind it. People want their thoughts in your head, just like they want their genes in your body. People want you to see something valuable and link it to them. My sister's friend, Lauren, has this big following on Facebook. She posted these recipes for healthy cookies, and I made a batch. Dates, almonds, and pecans – that's it. Delicious. Whenever I make them, I think of her. My wife just adapted the recipe to make a healthy pie. Again, delicious! Thanks, Lauren! We all want recognition. Ain't nothing wrong with that.

These are the ways that technology has created culture and provided us with a new playing field in which to play the genetic proliferation game. The combination of our resources and our technologies give us a value, GPP, which in turn, measures our ability to get a high GPR. The resources we already had, but the technology we developed. It provided such an advantage in getting high GPR that all humans who didn't get the adaptation of culture died. The Neanderthals are gone because they didn't adapt. Unlike our resources, our technology is not stored inside us, but outside us. As such it can be shared, it can be acquired, and it can be improved.

And so nature has built into us the drive to learn, to teach , and to always be on the lookout for new information that might give us an edge in life. 50,000 years ago we taught and learned about basic tools like axes and nets. Today we want to know what's going to happen with the economy, learn new ways to catch mates, and get rich quick. When we learn or teach, we get a happy feeling inside because we've done what nature wanted us to do. The same applies to beliefs and to the mental/emotional bond of the group. Nature made it feel good to believe in something, and we see so many people whose beliefs and the groups they belong to act as a great support system in helping them get through life. This is what culture is all about.

Information and beliefs are big pieces in the cultural puzzle, they've given us quite a lot. But there is one more big piece, and we wouldn't be here without it: Trade.

Trading

Just as we exchange genes to create new organisms that can be graded by natural selection, we now exchange two cultural resources:

1. Information – in the form of thoughts, beliefs, and instructions
2. Stuff – tools, extra food, etc.

Exchanging information and exchanging resources was the catalyst for our technological rise.

Trading is a big deal, biology-wise. It falls under the Fourth Recipe: Cooperation. All animals communicate on some level but, as Matt Ridley points out in *The Rational Optimist*, no animals trade like we do. I have 10 apples but can only eat 5 before they rot. You have 10 fish but can only eat 5 before they rot, so I trade 5 apples for 5 of your fish. Once people realized that they could trade surplus items, specialization became even more beneficial. You had one person who dealt exclusively with apples and another who dealt exclusively with fish. The more apples you could gather or grow, the more other stuff you could get through trade. This was an invitation for technology which, before this situation, was simply not there. We trade information and goods all in an attempt to improve our GPP. I can give you something that is not important to me in exchange for something that *is* important to me, and you can do the same and we both end up better off.

Most of us work for a company, so we trade our time and expertise for money. The company has money it's willing to part with, and I have time and expertise I'm willing to give. If I already have money, and they have the same time and expertise that I do, then we won't work together. That's Robert Wright's non-zero sum game again. All trade is based on this principle. If I think something is too expensive, I simply don't buy it. It only becomes worth it if I consider whatever I'm giving up to be of lesser value than the thing I'm getting.

Can you imagine your day without exchanging anything? Every day I give my time, skill, and money in exchange for other commodities. If I didn't and you didn't, if no one did, society would come to a screeching halt. But we all do it. We all want money, so we go to work. We all want food, clothes, a house, a vacation, a car, so we spend the money we made at our jobs. This works because I could never grow all the food I want, build a house, make nice clothes, or build a car and a plane to fly my family to Greece. Likewise, the shareholders of my employer could not do the work of 700 people. So we both have a problem, and we solve it for each other. **But the real beauty of trading is that we are each better off.** I can focus completely on specializing and my employer can focus completely on earning enough money to pay me. And so it was in early societies – you had one guy who raised goats, one guy who grew crops, another to take care of water, another for making weapons, a woman to make pottery, another to cook and make clothes, etc., etc. Before that it was "men hunting" and "women gathering and caring for the children," and there was no significant progress.

Trade is a real stimulation for innovation. When you trade, you give your extra stuff for someone else's extra stuff. So in order to trade, you need extra stuff. You get extra stuff by innovating. You do not get extra stuff by trying to do everything yourself. You're better off focusing on one thing and doing it well enough to be able to create surplus. For example, grow apples and nothing but apples so that you get so good at it and have a huge overload of apples. Now you trade those apples for the other items you need. Got it? And you will get good at growing apples. When you specialize in one thing, you have the time and focus to get better. The person who scatters himself around never really gets all that good at anything, and they won't have a surplus and won't be able to trade.

In a sense, when specializing, you become a subspecies of human beings. You are no longer just a person; you are a carpenter, a farmer, an accountant. You are a specialist and you can trade the surplus you make for all the other stuff you need. You have a huge advantage over someone who does not do this. You even see this in towns, like fishing towns and cattle towns. I understand there is a city in China that specializes in making socks.[19].

How Trade Still Translates to Increased GPR

Just to be sure you understand why we are innovating, sharing information, learning, and trading, let's do a little exercise. Imagine a man 1 million years ago who discovers he can put a few strands of leaves together and create a net to catch a rabbit that would otherwise be just out of reach. Let's call him Jimmy. And then imagine a similar man, Reginald, who does not make this discovery. Jimmy has a rabbit and Reginald does not. Jimmy can catch *more* rabbits by using the net. Reginald cannot. Jimmy's surplus of rabbits gives him a clear advantage in that he a) gains popularity with the ladies and b) gains popularity in the group. He'll have to share his newfound technology, and soon other males will have it too. But he is an innovator and the group will encourage him and others to make more cool discoveries. As for the innovator's genes, they're in better shape than those of his peers. His group's genes, which 1 million years ago were pretty closely tied with his own, are also better off. The potential (GPP) of the individual and the group are both raised by technology.

In today's world, if a man invents a spreadsheet, a smart phone, a Google, or a Facebook, then he is a hero. In practice, given the complexity of today's technology, people often work in teams. And given the complexity of the marketplace and the fierce competition within it, you can't just be an innovator, you have to be business-minded as well – but

[19] Datang: http://en.wikipedia.org/wiki/Datang,_Zhuji

business is also a technology. However you slice it, it still amounts to cultural success: using technology to improve your potential. The rewards technology provided 1 million years ago are the same rewards provided today. Why? Because, though we may have changed, the laws of nature have not. This applies to the individual and to the group. The individual is rewarded individually through improved resources, but is also rewarded by the group in terms of status, finances, mate selection, etc. The group, best defined today as his family, his country, his investors, or all three, are also rewarded.

All first-world countries have initiatives to stimulate innovation and technology. If you want to start a business, your country wants you to start a business too, and they'll probably go out of their way to help you. No country can have a good standing in the world without cutting-edge technology spread around to the most individuals possible (or barring that, huge reserves of oil). As groups exterminated other groups using technology millions of years ago, today's countries are in a similar battle to culturally defeat their neighbors. They might not be out to destroy them (that would be a waste of a valuable resources), but they are out to displace their beliefs and culture with their own. That doesn't mean they expect to succeed or that they don't sometimes choose the compromise of "live and let live," but in the current space of their subconscious, they strive for cultural domination. It's only natural.

Cultural Drives & Trends

Thoughts, beliefs, ideas, and technologies are the currency of cultural evolution like genes are the currency of biological evolution. I am trying to pass on my genes. I do this through impregnating women (well, just one) and helping to care for the children. In cultural evolution to achieve the same goal I do three things:

1. Pass on my thoughts in an attempt to gain allies in other people, raising my GPP.
2. Adopt the beliefs of others which I feel will raise my GPP.
3. Contribute to the group, preferably in the form of technology.

We can see memes and physical technology as outgrowths of genes. Ultimately memes are created by gene carriers (i.e., people) as an external entity there to help the genes get a higher GPR. The memes themselves: the information, technology, and beliefs, can have a positive impact on the individual, the group or both, but they are always attached to their source on some level. If that level is zero, then it would be a failed meme – one from which its source gained no real benefit. I use the term *source*, and not *creator*, since you don't need to actually create the meme, simply spread it. I mean, let's be honest, you didn't create your genes either, did you?

Let's look at a few examples of how these cultural strategies result in higher GPR today.

Everyone wants to write a book. Even **you** have considered it at some point. Why do you want to write a book? Go ahead and tell yourself whatever you want. "*I want to share my story,*" "*I have a great idea for a novel,*" "*Writing is therapeutic,*" whatever. I'm sure it's all true. But the underlying point is, you don't just want to write a book, you want the whole world to love it. Hell, I want the whole world to love this book! I'm not above adoration. No one wants to write a book that no one reads. Nor does anyone want to write a book and have the whole world read it and hate it. And no, no one wants to write a book that the whole world loves, but doesn't know who it was written by. . . .

So, why don't we want those things? As I mentioned, it's because we want to impregnate others with our thoughts and our identity. Just as I want to impregnate women with my seed, I want to impregnate men, women, and children with positive "John Khoury" memes. The more positive John Khoury memes there are in the world, the higher my GPP will be. Notice I said *positive* John Khoury memes. Adolph Hitler has as many memes in existence as Paul McCartney, but the difference is the positivity factor. Paul's genes will enjoy the benefits of his positive memes for generations to come. The man wrote "*Yesterday*"! How many chart toppers have you written, Hitler? That's right, none. Not only that, you were also responsible for the death of millions of people. And that is why your memes suck.

Pushing my beliefs and my ideas adds to my social network in a very incremental way – and I don't even have to know the person who has just been added to my social network. Cultural evolution gave a real boost to our social network resource. Modern society has enabled us to gain very small boosts to our social network by implanting our memes in the minds of many, many people – way more than the 200 total people we used to know in our tribe. The drive was the same back in the Sweet Spot – to make ourselves known well within our tribe. But the tribe has expanded to possibly billions of people, and we can now reach every one of them!

> ***"I was a fan of theirs before anyone had ever heard of them."***
> – one of my friends from college talking about some bandh

Today I shared a tune I was listening to with my friends on Facebook. "*Check it out, I'm listening to The Vines*[20] *(or whatever).*" I've always shared music with people. People love to share music with me. I share the movies, food, and TV shows I like. Why? Think back to high school. Didn't you have a friend who always "liked stuff before it was popular"? Don't some people want to almost *own* bands, like no one else can like them? I had a friend who matter-of-factly said he only likes bands *before* they become popular. I thought "what the hell – the music doesn't change just because they're popular." But now I see it's the ownership he's after. You want to have cultural products associated with you and when they become popular, that association is diluted to the point that you might as well not have been the one to tout it. You had something – a cultural resource – and it has been taken away. You associated with a person or group, they became a part of your social network, and now everyone counts them as part of their own social network.

---

[20] For the record, The Vines are awesome. See? I'm at it again.

After college, a different friend wrote to me: "*The Simpsons* and *Seinfeld*: always a reminder of you, friend." Every time he watches *Seinfeld* or *The Simpsons*, he thinks of me. I'm not *The Simpsons* or *Seinfeld*, but I became associated with them by constantly promoting them. And because their ability to get into people's heads in a positive way is stronger than my own, I can get farther by using them as a replacement for my boring self.

He came to mind as an example because he just died in a car crash. Devastating. What if it had been me that had died? He would have seen those shows and thought of me. Were we living close by, he might have called my wife and offered to cut the grass or go to the grocery store. Small stuff, but say it again with me: Small advantages matter in the long run to natural selection. And that's what makes the behavior stick. On the other hand, there are dozens of people who think of me when they hear Metallica. I never planned it this way. I was not consciously out to set this up, but our natural tendency to share thoughts and beliefs led to this situation. And what has it done for me? Nothing. Well, nothing tangible, I'd say.

But remember that our actions today are based on the results of the Sweet Spot. If, back then, we received a slight advantage from this behavior, then we will continue to do the same sort of thing today, even if the payoff is slightly different.

Build Up Your Team by Sharing Information

"I want everybody to go to my funeral."
– celebreneur Gary Vaynerchuk

Human beings share memes to gain allies. It happens every day all the time. This is all commercials are. Yes, they are also trying to sell you something but most marketing departments of big companies understand that they are not making a direct sell when they show you commercials – this is why *sales* and *marketing* are different. Took me a while to understand that. Sales is signing contracts and the actual exchanging of goods and services for money. Marketing is *branding*! What an appropriate term. Corporations are imprinting their identity on your brain. Sometimes you are aware of it and you turn off the TV. You subconsciously know what's going on and you stop the process.

As a meme, religious doctrine seems to be on its own level. Sure, we like certain brands, certain artists, certain ideas and subjects . . . but people *love* religion. Just as the sense of smell is tied more closely to your brain than other senses, religious belief must be more tied to the brain than other types of memes. Music is a close second, and note how important music is to any religious ceremony. But somewhere in our evolutionary history it became advantageous not just to believe in religion, but to do so fervently, as we saw with how the zeal of two different groups to fight was so affected by the ceremonial rituals they regularly held or didn't hold. This being the case, pushing religious belief has been a very effective way to gain allies. Nicholas Wade, again in *The Faith Instinct*, likens religious belief to the acquiring of language. You learn it at a young age and it becomes a part of you. It's all so personal, so intimate and so close to the heart of life, that it opens people up and allows that sticky part of their brain to glue itself to you and all that you are doing.

This is also done on a group level. Religion makes for an obvious, relevant example. In *The Evolution of God*, Robert Wright outlines how early religions were used as ways of merging a group of defeated people with their new overlords[21]. Group A defeats Group B in a war and Group A then tries to get Group B to gracefully accept their new rulers. This is a slight progression from when Group A would kill off all the men in Group B and take all their women as sexual slaves. That worked when

---

[21] Yes, the same religion that has so often denounced evolution is also a product of it: http://www.quanology.org/robert-wrights-the-evolution-of-god

groups were smaller, but once you reach a certain threshold in numbers, you're going to need some complacency from the losers. Besides, killing off all the men means they can't work for you later, and can't help you fight in your upcoming battle with a bigger group. Religion was there to grease that wheel.

If you look into it, you'll find that many Christian and Jewish traditions actually come from Pagan ones. Christmas on the Vernal Equinox? What a coincidence. . . . Christians are quite happy to suppress the idea that Jesus was not really born on December 25th. So, why do we celebrate his birthday on that date? It goes like this. Christians would rub shoulders with Pagans and then merge bits of Pagan beliefs with their own in an attempt to get Pagans to convert to Christianity. It's necessary to ease the transition. As Nicholas Wade notes, you don't just destroy their temples and replace their holidays with new holidays on different days. You re-use the same days and give them a slightly different twist. You leave their temples, but remove the statues and tidy up a bit. Give it all a little time and they'll become Christians. Pagans adopted a new belief system and were better off for having done so; they were able to hitch a ride on the Christian rise to prominence in the Western World.

Today the same phenomenon exists: we adopt the collective beliefs of groups which are good for our genetic career. Those groups might be religious, they might be political, they might even be sports teams. Mostly, we do this without consciously knowing why. It's no wonder the Yankees have more fans than any other baseball team. Would I have fallen for the Beatles if they weren't the most popular bands of all time? And it's no wonder people choose the popular religions of their home country. If you didn't have Muslim parents, you would be foolish to *choose* to be Muslim in America. Likewise, it would be foolish to choose to be Christian in Indonesia if you didn't already have Christian parents. It's not that one religion is more believable than the other, but, again, belief adoption is not about the credibility of the belief, it's about the boost in GPP one gains from joining its organization. So many people join religions and go to churches and synagogues because of the support it provides them. They are often very up front about it. And this is why organizations form around belief systems. An organization — similar to a good electronics manufacturer — is a machine there to exploit the potential of a growing belief and create the level of support that people want. It's like creating an organism out of a growing heap of unified cells. Just like a company has to be organized and professional to sell products on a reasonable scale, a belief needs an organization to increase the GPP of the belief holders to the highest degree. Not organizing would be wasting opportunity and that's just not done in nature. So, complain all you want about organized religion – there will always be "organized something." It's a part of nature. But I digress. . . . The point is that we are constantly offered beliefs, and we accept or reject them based on what they mean for our GPP, and not on their credibility.

I would now like to confess something to you, dear reader. We've been together for a while now, and I feel like I can really trust you with this. . . . I . . . I was a young Republican. I didn't join some committee or anything, but I was on the same page as the conservative thinkers of the mid '90s. That's what you were if you were a young white male in my circle of friends. I would listen to radio shows and nod in agreement as they told us what was so horribly wrong with America and what needed to be done to fix it. I would have stayed that way, I'm sure. But in 1996, at the age of 22, I moved to Holland. The Dutch do everything differently and look at everything differently than Americans do. At first I thought it was all stupid, as one often does when thrown into a new environment. Slowly but surely, I started to agree with the things they were doing, the ways they were thinking. And before you knew it, I understood and supported social democracy.

What the hell happened!? Were my beliefs that malleable that all I had to do was move to a new country? Did that mean my new beliefs were potentially just as worthless as the ones before? I can't say. What I can say is that it's natural to adopt the beliefs of your environment, just as you adopt certain behaviors and manners of speech. Give me a few days in England and I'll be talking like the Queen. A few days in Brooklyn and I'll be talking like Kevin McGovern, ignoring *r*'s at the end of words. It's no wond-ah that so many children adopt the religious beliefs of their parents and/or environment.

Beliefs are incredibly dear to humans. They are our children in our cultural families. It should be clearer to you why there are only two things in the world people would die for: their children and their beliefs. People feel out of place when they are not allied to a belief system. This has been my biggest complaint as an atheist. I want to be part of a group too – not just a bowling league, I mean a group bound together by shared beliefs in something deep and meaningful. Being an atheist means being cast into a sea of cultural chaos. Religion is nice and structured and its drive to include so many people means it is always a compromise – hardly anyone believes *everything* that is professed by a religion.

I'm not putting people or beliefs down. I'm simply trying to shine a light on their true nature – what they're here for and why we are so militant about them. Breaking away from the beliefs of the group is akin to breaking away from your parents. For a child this means certain death. In early civilizations, when you no longer believed what the group believed, you were on your own in the same fashion. It was just not done. And that's why the impulse to believe what your surroundings believe is so strong – strong enough for the believers of today to deny the phenomenon completely. As fundamentally as genes and cells make us up physically, beliefs and thoughts make us up mentally.

Cultural Evolution

. . . And then the fighting starts.

We see that thoughts, beliefs and technology can be seen as similar to creatures that evolve. One day there exists a set of technologies and ideas and, after many generations, another set evolves to take its place. Competition is the push for this evolution. Competition is a fixed element of the universe just like matter and gravity. We would be well-served to love it just as much as any natural phenomenon.

As animals, we are in competition with other species for resources. As individual people, we are in competition with each other. As groups, we are in competition with other groups. Fortunately, we won the race against other animals. There was apparently only room for one winner and that was us. We crossed over onto a new plain, leaving the old methods of biological evolution in the dust. It was like a leg race and there was one car for whoever was lucky enough to get into it. Once we got in it and hit the gas, we left the others behind.

Competition between human individuals and competitions between groups are, however, still in effect. They are not biological, they are cultural. We compete technologically/culturally and the individuals and groups with the best technology and culture win the competition. If there will be an all-out winner in cultural evolution, it won't be anytime soon. It's still evolution and it's still slow. But the point is that technology, as we know, has no end. It is always developing, and always competing via the people who use it.

I want to stop for a sec and talk about numbers again. Let's talk bottom line. We each want a high GPR. So, why don't we all just bonk like rabbits and have 30 kids in each generation? That'll up your GPR! But anyone who has kids knows that they are more work then you could have ever, ever imagined. Not just financially, but in terms of total energy spent . . . and that's what life is all about, right? Energy. Even 100,000 years ago, kids were a pain in the ass. This is why they are so damn cute. If they weren't, you would kill them, or at least leave them for dead.

Human infants are helpless little creatures who wouldn't last an hour in the real world without constant care. How this predicament came to be is another story, but suffice it to say somewhere down the evolutionary line, natural selection chose for high-maintenance children. This is also the case for most large mammals.

Essentially this choice puts us on the low-risk/low-return side of investment strategies. Why? Because low risk requires higher quality human beings – those more able to adopt technology and leverage its possibilities. If you only get a few chances to do something, as we do with our few offspring, you need to make sure you are in the best shape possible to succeed. You can't just hope that the next kid will pass on your genes, because there's a good chance there won't be a next kid. Therefore humans must be of high quality. Relative to other mammals we are pretty far on the low-risk/low-return end of things. And these days, even more so than just a few hundred years ago. Think about it, we could have 20 children and not take care of them. Each one would be a GPR stab in the dark. We would not expect them to survive or even reproduce, let alone create tools. This would be a risky approach. But if they all did well, by some stroke of luck, we would be set. I mean, there's 20 of them! The opposite strategy would be to have like two kids and put tons of energy into raising and nurturing them. They would have a much better chance of surviving, reproducing, and passing the Bar Exam. Much less risk. But, yeah, there's only two of them. All we can say is that nature chose the second strategy for us and we are most rewarded when we take care of our children. It worked out best in the long-run. As it turns out, the "long-run" is a solid strategy for genetic proliferation.

How do we create high quality people? It's the same today as it was 100,000 years ago – you educate them, and improve their capacity to do more and take on more responsibility in the world. Educated people are effective people. Early humans needed to make sure that work got done. Low risk, right? That means you need to succeed in fewer attempts to do whatever you're doing. That is what culture provided – a higher success rate. We are a species of teachers and learners. The more we taught and the more we learned, the higher our success rate became and the more we moved down the low-risk/low-return line.

This has been the trend for as long as we can remember. Our culture and technology made groups and individuals more robust and successful. Of course, as one group developed, other groups developed too, albeit at different rates. Whereas groups were physically competing hundreds of thousands of years ago, now, in a sense, their cultures and technologies became competitors. Hand axes would compete against spears. One would win a battle but the loser might be incorporated into the cultural arsenal of the victor. Cleaning up left-over weaponry from your foe is a great way to stock up for Christmas. Today, big companies buy smaller companies to acquire their technology in the same vein.

There are two main advantages cultural evolution has over biological evolution

1. It can be passed on **within** a generation. I can learn something today and teach it to others tomorrow. Compare that to the average 25 year waiting period for genetic improvement.
2. Many individuals can be involved in each improvement

This being so, the development of ideas and thoughts often progresses very quickly. Now a-days, we experience major cultural changes every 10 years. I have no idea what kind of crazy things I'll see by the time I'm 80, but I'll bet it would blow my 38 year-old mind today. We can take this for granted. 10,000 years ago, not much changed in the course of a person's lifetime. No one looked at a hand axe back then, scratched his head in wonder and thought, "What'll they think of next!" There was no "next." The amount of change experienced by every subsequent generation has increased exponentially and continues to increase. Behind every change is a new technology, in the form of thoughts, ideas, beliefs or strategies, and these technologies are in competition with one another as if they were independent entities – just like genes. And just like genes, they are constantly being judged by natural selection. Only the "good" ones survive. In this sense, yes, some are better than others, just like some adaptations of animals worked while others didn't.

The downside of cultural evolution is that it's less closely tied to genetic proliferation than biological evolution. In human biological evolution, a new baby means a 50% success rate of getting my genes passed on – I get 50% and my mate gets 50%. I have to have two kids to get 100%

of my genes passed on, but this only results in a total GPR of zero since I invested 100% of my genes to have those babies. Sounds like crap, but it's the most bang for my GPR buck I can get. In cultural evolution, a new technology or idea will help my GPR but not as greatly since its impact is spread across multiple individuals, thereby diluting the effect it has on my genes alone. But the impact is still there, and as new ideas and beliefs are created more quickly and easily than offspring, our genes are happy and they make us feel fulfilled.

Humans are in a cultural arms race, whereas other living creatures are still in a biological one. It may be a taboo to say that some cultures are better than others, but that is the inevitable result of this fight on the battlefield of evolution. I'm not passing judgment, I'm just reporting on the judgment that natural selection is passing. Groups of people are not so easy to classify anyway. It would be overly simplistic to say "all Arabs" or "all Africans" since there are groups within groups with very different cultural practices. But as we look at the world and see how the various groups of people are faring, it's not hard to conclude that some are on their way up and some are on their way out.

I would argue that groups are not conscious, only individuals are. This implies that countries arrive at a state of disaster or greatness as a result of natural circumstance – what Germans call the Zeitgeist. China is doing well because it was in the cards. America did very well by the same wind of prosperity. Syria is going through a really rough time because of forces which are beyond anyone's conscious control. In all cases "some person" just didn't grab the steering wheel and consciously lead them in a certain direction, rather millions of people unconsciously steered in one direction over a long period of time.

What we need to take away from this idea is that nature is in full control. Some guy was asked if he could ask God one question, what would it be? He responded with "*Is the universe good?*" From this perspective, we see that the answer is yes. Natural selection only allows memes to stay which progress man-kind. This is where I stray a bit from some biologists – I don't see that memes and genes have lost their link. If a meme does not serve the genes using them, they go away slowly with time. Even cat pictures that do not serve their posters will go away . . . but they *are* serving their posters so they are going *nowhere.*

We see the same trend with foreign policy, which we can also consider to be a meme. Being aggressive used to serve countries well. Slowly but surely, aggressive stances are being replaced by more open ones. Hundreds of years ago, aggressive countries like Holland and Sweden refocused their effort on trade and won big. War memes lost to trade memes and the peoples of those countries did well. They made the switch not for moral reasons, and not consciously but only because war was no longer working. They had no choice but to find a new set of memes to latch on to. They didn't make the decision to change their ways, nature made it for them. She's making the same decision for new countries one by one. By now we understand two of the levels of evolution (oh, yeah, there's another one coming up) and how they have affected our species biologically, and then culturally. But what about us as individuals? Are we just along for the ride while evolution is at the wheel. Are our minds slaves to the drives of our genes? Are we just unexceptional vessels devoid of free will? Hardly. We've got personality.

Life Strategies

So, here we are *living*. Now comes the question of how best to live life? We all have some innate knowledge about how to accomplish basic biological tasks like chewing our food, breathing, sleeping, and even how to have sex and care for our young. But being the cultural beasts we are, how well we perform in the cultural realm, can have a positive or negative impact on our lives. The group dynamic is at the foundation of culture. You are an individual who must fit somewhere in this group. How you fit and how that fit leads to your GPR is manifested in your personality.

GPS – The New Personality

Your personality is who you are. I remember in Intro to Psych wondering why there was this separate field of psychology studying personality. I thought everything was personality. It's not. Personality is your essence. It forms in an early age, it has a big genetic component, and it doesn't change. Things like behavior and cognition are layers above this. I can learn to be more assertive, I can decide to work out more, I can think two ways about the same thing, but ***I*** will always be the same guy. Get it?

And who is that guy? What makes us who we are? Psychologists boil it down to essential character traits that we all have whose levels don't really change much throughout the course of our lives. Let's take a look at the Big Five[22]:

- Openness – characterizes people who are creative, curious, and insightful
- Conscientiousness – characterizes people who are self-disciplined, precise, and deliberate
- Extraversion – characterizes people who are energetic and enjoy interaction with people and things outside yourself
- Agreeableness – characterizes people who like it when we all get along and are sensitive to the feelings of others
- Neuroticism – characterizes people who are moody and experience a fair amount of negative emotion

We can think of these five traits as knobs on a machine. We each have the five knobs but we all have them turned to a particular setting. Your personality is the sum of your levels on each knob, which remain relatively constant. A person who is pretty easy going as a child will largely be so throughout his or her life. Someone who is pretty neurotic, like me, can try all kinds of ways to be less neurotic, but will ultimately fail. I know, that sucks, right? But it is what it is. Nature allowed these traits to be in

---

[22] This is just one characterization used by psychologists. There are others. http://www.quanology.org/personality-characterizations/

humans because they all work, so it's tricky to say that a certain level of any of these traits is better or worse than another. They're all strategies that have worked for a long time.

Of course, the other major factor to influence personality is the way your parents and family treat you early in life. This explains why identical twins can be so different. They are forced to turn their knobs to opposite settings. One turns their agreeable knob to *nice* setting and the other one turns it to the *jerk* setting. This is a basic phenomenon of all social systems: similar strategies will repel each other, so people take on complementary positions rather than fighting for the same space. This general phenomenon is called *Adaptive Divergence*[23]. If two siblings that live together are close enough in age, they will be different types, despite having similar genes. This *personality divergence* gets its initial direction from the genes and, as early on as it can, starts to read signals from its surroundings. This happens in twins, genetically disparate siblings, in couples, and even in animals.

All higher animals have personality. The notion that animal personality is different from what we humans have is rather silly. Anyone who has spent time with two different animals of the same species knows that they have unique personalities. You can expect a certain pattern of behavior from one and a different pattern from another. Like, one dog will have a hard time listening to your instructions and another can easily be taught to fetch the damn newspaper and drop it at your feet. Humans just have a more diverse range of personalities, and we are better able to communicate and express them.

We've talked about how technology is at the heart of culture. We've also talked about individuals and their value (GPP), which is the sum of their resources and technologies. Now, we are going to talk strategy. And to reiterate, strategy is a technology of *process*. Strategies for people boil down to a pre-programmed way of doing things in a group setting.

---

[23] Adaptive Divergence from Frank Sulloway: http://www.oxfordscholarship.com/view/10.1093/acprof:oso/9780195372090.001.0001/acprof-9780195372090-chapter-4

Your strategy for operating in a group with the goal of high GPR is your personality.

I have put personality in the context of evolution by coining the term GPS: **Genetic Proliferation Strategy**. There are many different kinds of people who can mostly be grouped into a handful of different archetypes. These archetypes encapsulate the GPSs of their owners. The hero, the leader, the divine beauty, the caring mother, the asshole, the joker, the whore, the playboy, and the devoted dad are all just strategies we were assigned at an early age as a way of fitting into society in our quest for genetic proliferation. Animals have GPSs within their species as well. Gorillas have a dominant male, subordinate males, loose females, and "loyal" females. Which one you are shows up early on in your life. It is determined by genes and early environmental circumstances, and is immutable. You are who you are.

Now, in the course of writing this book, I've bounced some of this off friends and have gotten some strange responses. "We are who we are because of how we were raised." I remember this too from my Psychology classes. Indeed many text books have been written explaining how experiential learning makes us who we are. "Men are destructive as a way to get attention they missed out on as a child." "Girls like dolls because their parents actively promote them as toys in attempts to make girls soft in nature." "Babies cry when they learn their parents give them attention when they do so." Is this all true? I don't know. What I do know is that much of our current behavior is a direct result of the evolutionary benefits of that behavior. Some behaviors are more deeply rooted and therefore more universal and inflexible while others are more easily adapted to modern life. Things get sticky when we think things like boys rough-housing is something they can unlearn. Such things are primarily rooted in biology.

My middle boy learns that when he does something tough, his older brother is impressed. But his older brother is only impressed because it is tough. Life in the Sweet Spot favored toughness, so those exemplifying it were appreciated. If my son were cuddling a doll, his brother would not be impressed but rather repulsed. Learning is always in the direction of higher GPP. We learn those things that take us to better places. He acts tough and his brother gene-ishly likes it because it gives both their genes a better chance of success in a tough world. I might not like it when he acts tough, but my reaction doesn't lower his GPP as much as his older brother's interest raises it. In fact, the only real reason I don't like it is because he is breaking things in the house, thereby lowering my GPP. Otherwise, I'm fine with him being tough. Learning is more often than not, fine-tuning the things that nature wants us to learn.

What elements of our modern society are learned? To fight and make wars, to have sex, to love our children, to stimulate assertiveness in men and beauty in women? No, the basic behaviors we see in all cultures are not learned as behaviorists would have us think, they are instinctive – as instinctive as a lizard who comes out from hiding when the sun starts to shine. Boys learn to be tough and girls learn to be pretty. But those drives have been there for millions of years, so we cannot say that they are a product of modern society, nor can we turn them off by restructuring society.

We are flexible enough to allow certain behaviors some slack to move to the left or the right as we maneuver within a cultural world. We learn stuff along the way. We seek positive reinforcement and avoid negative reinforcement. But we are not a blank slate who can be taught to be entirely different from what we universally know to be human. Our personalities, the true essence of who we are as individuals, are biologically rooted and won't change with a couple of new insights here and there.

When I mentioned this to a friend, he was not cool with the idea of being immutable. He is a loving father and husband, a productive member of society and an upstanding member of the community where he lives. But he was not always this way. He told me a few stories about going out and getting drunk and kicking in car windows. He once put a hamster in a dryer. In his defense . . . no, there is no defending putting a god damn hamster in a god damn dryer! And kicking in car windows? What kind of hoodlum . . . The point is, he changed his ways.

If personality is fixed by biology and early environment, then what turned this reckless vandal, this animal-abusing monster into the wonderful human being he is now? His hormones. Your essence is constant, but your changing hormones and situation can have an impact. When we are young and competing for mates, as I've explained earlier, our hormones are different because our goal is different. We want a mate. And our behavior is tweaked by our bodies to drive us and help us to achieve that goal. And so we go through phases where our behavior is more directed to the goal of that phase, but our personality remains fairly constant underneath. As we've seen, our main goals are mating and parenting. Though they go together, they are very different in terms of what they require from you. As such, your behavior will be very different when trying to find a mate and when trying to raise kids. What will not be different is your general level of agreeableness, your openness to new things, your level of neuroticism, etc. These traits are more hardwired than your behavior and so we consider your personality to be a fairly fixed set of dials created and set by your biology and then tweaked in the first few years of your life.

Having a large biological component means something important for our personality. It means: a) Our personality is bound to the Sweet Spot, and b) It is geared toward biological drives.

Understanding the implications of this is the key to understanding human nature. When studying a wild deer's white tail, scientists conclude that the purpose of that tail is to signal other deer when danger is near. If you put that deer in a zoo, you have removed him from the process of biological evolution. He no longer needs the white tail to signal danger because there is no more danger. But it took millions of years to develop that tail, so what you cannot remove is the behavior associated with that white tail. No matter how many zoo transfers he'll make, no matter how much evidence he can muster that there are no more predators stalking him, it is still in his nature to raise his little tail when he feels threatened. We are in the same predicament, and understanding our nature depends on understanding the pressing drives and motivations of humans in the Sweet Spot, before we moved into our little zoo.

Sexual & Parenting Strategies

"How do I get that girl to come home with me?" and "How do I get that guy to notice me?" will both involve sexual strategy, but before the details of these become clear we have to talk about the gene's sexual strategy, not yours.

Quantity vs. Quality

I've got 5000 bucks. Quick, what should I do with it? I should invest it, right? OK, so I lied about the 5000 bucks. But I have something else that's valuable: *me*. What should I do with me? What do my genes want me to do?

Let's talk investment strategies for a minute. If you're wealthy, you know about "investment strategies." This is how you handle money, that valuable resource you have. You want it to grow. You want to "make your money work for you." That means putting it places where it will get a high return. Because if it just sits there under a mattress, inflation will make it less valuable every day. So, where do you put it? There are lots of options, but all the options fall somewhere along the gradient of high-risk/high-return and low-risk/low-return which we mentioned earlier. If you've ever played the stock market, you know that stocks are usually higher risk. That's why it's called "playing" the stock market. Investments in businesses are slightly higher risk than stocks. Gambling is the highest risk you can get – the lower the odds the bigger the payoff. Mutual funds, government bonds, real estate are all lower risk, but they don't have the big payoffs that the high-risk investments potentially do. And that's how it goes – the higher the risk, the more return you expect if your number hits, and the lower the risk, the less of a return you can expect. A high-risk/low return investment will soon be extinct like a cheetah who tries to swim after tiny catfish.

Back to people and genetics. Ever see *Idiocracy*? It's a movie by Mike Judge, the maker of *Beavis and Butthead*. It shows an interesting view of the future. In the beginning, we see two different couples, one well-educated and conservative, the other low-class and reckless. The conservative couple is debating about having children. "We can't have them now because we're so busy with our careers, blah blah blah. . . . " They only want two children anyway. The other couple, some trailer park hillbillies, just get busy whenever the mood hits them. The guy actually knocks up his neighbor. His eldest son shows the same sexual promiscuity and bones everything he sees. Children abound. . . . Sure, this is fiction, but what does this dichotomy show us?

GPR is a number and the same investment framework used for money can be used for GPR. On the one hand, you can pass on a lot of your genes and invest very little energy in their maintenance and in doing so increase the risk that your offspring will not survive. Or, on the other hand, you can pass on fewer genes and invest more energy in their maintenance, lowering the risk that they will not survive to pass on their subsequent genes. Frogs, for example, have about 1000 babies at a time (I've actually counted), but they invest no energy in them after the eggs have been laid (which is how I was allowed to count). As we would expect, very few survive (it wasn't my fault). Elephants, on the other side of the spectrum, give birth to one baby at a time, and both parents invest a good bit of energy in raising the baby to a self-sustaining level.

What about those two couples from *Idiocracy*? As I said, you have to think of "people" and our motivations as they were when we stopped evolving biologically, back in the Sweet Spot. Obviously, we are more like elephants. We have one baby at a time, and most people have fewer than 10 children in a lifetime. So, relatively speaking, we take the quality route. Frogs, on average, are involved in several thousand tadpole births, out of which maybe 10 live to sexual maturity, and people, traditionally, are involved in about 4-6 child births whereby about 2-3 live to sexual maturity. It's pretty comparable. Within human sexual behavior the high-risk/low-risk ruler can be applied again. Some people prefer to settle down and have fewer children, others are a bit more "wild," i.e. promiscuous, and have more children but invest less energy in their upbringing. They are often called cads, dead-beat dads, etc., but hey, it's a strategy, and if it works then nature lets it stay.

We see all these traits in the personalities of the people we know and love. Some men and women are more adventurous and some are more reserved. The adventurous ones have a harder time in relationships and maybe even jobs since by definition their personality is geared to higher risk mating strategies. Love 'em and leave 'em if you're a guy. And "falling for the bad boys" if you're a girl. The reserved folk will choose a life of less expectation. They seem content with what they have. They will have stable relationships and jobs as is dictated by their low-risk approach. This even happens on a group level when poor societies are subconsciously forced to take the high risk approach. Their citizens have many children since they expect a good number not to make it, or at least not to do well in life. Once their society develops and requires more advanced technologies from their people, they have fewer kids and focus on the quality of their upbringing. This shift is a biological strategy for increased GPR.

Most of our choices in life, from an evolutionary point of view, are sexual strategies. This doesn't mean we're only out to have sex, or that we're only out to raise children, or are even consciously aware of any of this as we make the routine decisions in our lives. What it means is that GPR is the main drive and everything we do is in service to that drive. It is through modern technology that, for that same goal, our day-to-day lives look totally different compared to the Sweet Spot.

Like any other group of animals, we compete for scarce resources. And just what were the resources in the Sweet Spot? Two big ones: food and . . . wait for it . . . eggs. Female eggs are a resource to a group of mammals like us. Without them, we cannot reproduce. Starting from around age 12 and until age 45 (and many people did not live to age 45 back then), each female only creates about 12 eggs a year. Male sperm is not at all on the same level. Men produce way more than they could ever use and continue to do so into old age. This is why human society has always been male-dominated. Despite the thought that women hold all the cards since they are the ones with the scare resource, it is males who dominate because of their drive to get this resource. Women don't have to compete as fiercely for male sperm – just go to Ladies Night at your local bar to confirm this statement.

It is men who have embarked on an arms race to get this scarce resource. While life in the Sweet Spot was fairly egalitarian – especially when compared to medieval Europe, men, as we've learned, only had to take a small risk for the potentially big genetic payoff of mating with multiple women. And while women could always get extra men, the cost of doing so was far greater than the potential reward. When a man impregnates an extra woman, he has significantly increased his GPR. When a woman finds an extra father to have a new baby, she does not get the same reward. Why not?

Think of computers. Early PCs could only perform one operation at a time. This was called linear processing. And then along came the technology of threaded processing where computers could do multiple operations in parallel. You can launch Word and Excel at the same time and your computer executes those two requests in parallel. That is what a skirt-chasing man is doing. Wanting "a little on the side" is the manifestation of the male drive to increase his GPR through parallel efforts. Women, on the other hand, can only have one baby at a time. Sure, they can have twins, but there was no strategy for doing so in the Sweet Spot. The reward that a man has for having 5 affairs at a time is not there for women and therefore women did not develop the same drives that men did.

For much of our evolutionary history, some males got a disproportionate number of women. That is to say, they impregnated more women than other men through adulterous strategies. Our modern world has allowed this trend to blossom, so some guys get lots of women, while others get only one, and still others are at home alone. Let us remember that while some men are out impregnating more than one woman, other men are getting sand-bagged in cuckoldry. **Not all men got to be fathers back in the Sweet Spot, but all women got to be mothers.** Men had more to gain by being promiscuous but also more to lose by not being so. Add this to the fact that men were the predominant victims of murder and you can understand why men are edgy, aggressive, and constantly scheming. Ladies, your man is emotionally distant because he is trying to figure out how to make it in this world, 'cause it's not a given that he will!

But that's not us anymore, is it? We're not constantly under the threat of murder. And monogamy is law, which means one girl for every boy.[24] But, just like with the white-tailed deer, removing us from the biological reality of the Sweet Spot does not remove the drive or the behavior. Human nature is still largely a result of our drive for genetic proliferation. What did we need to proliferate our genes back then? As you can imagine, it is very different for men and women. For now, we will focus on the ladies, since their situation determines male behavior.

The Female Arms Race

I feel like I have been overtly focused on male behavior in this book. Yes, I'm a man and I find the male "arms race" I mentioned incredibly fascinating. It's done wonders to help me understand myself. But if I take a break from it to look at women, their behavior and drives are just as fascinating, and equally offensive. We often hear that "men are pigs" and I believe that, often enough, we will admit it. We don't like being pigs as much as women don't like us being pigs. You think I like constantly checking out women? I dream of being able to walk hand in hand with my wife down a pretty city street filled with short skirts and tan legs and not be at all distracted. That which annoys my wife annoys me too. But women are no angels either. What's more, "How dare you even suggest otherwise!" Women like to keep this fact well tucked away. Whereas male nature is out in the open, females keep theirs deep in the

---

[24] Yes, it was men who pushed that law through their local government, not women.

closet, behind last year's shoes. Why don't men call women out on their B.S.? Because then they would lose favor with women, and as we know, they are a scarce resource, so the vast majority of us need to stay on their good side.

I love women, so I'm not out to get you, just out to get the truth. It may hurt, but we'll all be better off knowing it in the end. When it comes to GPR, women are just as competitive as men. But because they don't experience that extreme difference between "winning and losing," they don't have the aggressive strategies that men have. There's a huge gap in terms of what's at stake, but women still want to win. Just as men need women to reproduce, women need men, too. We need their eggs, but they need two things from us: sperm and support.

As we know, sperm is cheap – even good sperm. Just check out the price differences between sperm and eggs at your local fertility center. Women get to be choosy. Quality men by definition are a limited group and they get a disproportionate amount of attention from women. Since these quality men have endless sperm, they are not exaggerating when they say, "There's plenty of me to go around, ladies." And if a woman were truly only interested in sperm, there would be no reason for her to look elsewhere. But as we know, women want more. It is a tough world, man, and raising a kid takes a whole lot of time and resources, and well, a girl is going to have a hard time raising a bunch of kids on her own. She's going to need some support. And through the course of human evolution (and this counts for most large mammals) the strategy has developed where men stick around and help raise the kids that they fostered.

Why? Why stick around and raise kids when you can just have a baby and start work on the next one? Because it increases the chances of high GPR. How many of those kids did they expect to survive to produce kids of their own? Maybe half? So without the support and protection of a man, that number is even lower. Women have had to make men prove that they would stick around to help raise the kids. A man invests less than a woman in conceiving a child (10 minutes of pumping compared to 10 minutes of lying down, nine months of pregnancy and hours of painful, painful labor), so he's not as invested as she is. But she needs him to be invested in order to keep him around, so what does she do? She gets him to invest more. She promises that life together will be heaven – not by telling him how great *he* is, but by showing him what a devoted mother and spouse *she* will be. At the same time, she tries to get him to actually invest: give her stupid things like diamonds, flowers, and expensive dinners. With these things he *actually invests* and ends up bringing himself to her level of investment so that he is also less willing to leave later. Some types of men stick around more than others and some types have more to provide than others. Ideally a woman scores a man with good genes, lots of resources, and the desire to give both (especially the latter) completely to her and her children. These "good men," as we all know, are hard to find. So a woman compromises.

> ***"You know the number one reason why your woman is pissed off? . . . It's 'cause you're not her first choice."***
> —Chris Rock

I want to stop briefly to justify the above and what's about to follow. These descriptions of male and female behavior tend to make both look like subconscious animals; like dogs just doing what nature demands of them, completely void of conscious thought, free will, or morality. The people I describe are repulsive, no different from the vermin we so eagerly remove from our dwellings. But we do need to suspend judgment to get to the bottom of things. We have to go along with the fundamental idea that our genes want to reproduce, as they have been doing for billions of years. And those genes are the ones behind the wheel every day of our lives as we go to work, find a mate, and love our children and family. The numbers are too compelling to ignore. 3.7 billion years of control will not end because humans start listening to hip-hop. We must consider the idea that nature is in control, as always, and that humans are subservient to nature. And if we don't like it, our best chance out is to understand it and then create a strategy that shifts the control into our hands.

Right, let's continue where we left off 50,000 years ago, with women, you little she-devils. It's a jungle out there and you want a strong man to pass on his strong genes to your children. Women want Rob Lowe with his square jaw and symmetrical features, which are signs of gene quality and dominance. They don't want Ricky Gervais with his short, pudgy stature. But then, Ricky has done awfully well for himself these past few years. I mean, Rob probably isn't starving, but Ricky doesn't know what to do with all his money. . . . What if . . . What if a woman could sleep with Rob Lowe, score his genes, then sleep with Ricky and claim the baby was his? That's just crazy enough to have worked . . . for millions of years. . . . As we've said, it's called cuckoldry. Google it. A woman gets the genes she wants from one man and then gets the resources she wants from another man. All she has to do is trick Ricky into thinking the baby is his.

Not convinced that women do this? I know you would never do this; I'm talking about statistics and extremists here — women 50,000 years ago — not about you good girls who would never consider such a thing. But let's look at the evidence. Yes, evidence can come in the form of other animals. We are animals and if certain behavior is prevalent in our lives and those of another animal, we can study that other animal to come to conclusions about ourselves.

Let's talk again about cuckoldry. Birds do it all the time. They appear to be comparatively as adulterous as we are. Ridley points to a study showing a good 25% cuckoldry rate in swallows. Still not convinced? What about our ape cousins?

Testicle size shows us a few things about mating styles in animals. The average size of the testicles in males of a certain species is related to how much sperm competition they have to deal with. What's sperm competition? You don't know? It's when the sperm of two men residing in a vagina are fighting to see who actually gets to fertilize the egg. That begs the question: What are they both doing in her vagina at the same time? Hhhhh . . . Hate to break this to you, dude, but she's sleeping around. But don't worry, you may be heart-broken, but your testicles saw this coming and therefore created lots of sperm so as to drown out the other guy. Large testicles for the males of the species imply that the women are sleeping around.

Thankfully, human testicles are not that big. They are smaller than a chimpanzee's – their females get around. And they are larger than a gorilla's. Male gorillas watch over their harem and are quite sure their ladies are not getting any on the side, so when his sperm travels through her vagina it's like a lone car driving up I-95. Still not convinced? Ridley points to studies by Baker and Bellis that show that women having adulterous affairs, through orgasmic manipulation, are twice as likely to get impregnated by their lover's sperm than by their husband's sperm. Still not convinced? Baker and Bellis also showed that men ejaculated twice as much sperm when they had spent the day away from their wives as compared to when they had been with her all day, as if the testicles are assuming there will be competition in there since they have not seen her around. It's all subconscious, of course, and the man may not even be aware of it, but evolution is. Still not convinced? Watch that Woody Allen film *You Will Meet a Tall Dark Stranger* and you will see Anthony Hopkins' new bimbo girlfriend getting pregnant. He asks her, "How can I be sure it's mine?" After some discussion and admission of infidelity she breaks down and says, "What's the difference? You always wanted a kid."

Of course, that last point shouldn't convince you – it's just a movie. I'm just trying to get you to have a look around at the people you know and the things you've heard. Women want two things very badly: a man's support in raising her children and good genes. They want to get the best version of each of these things, but the best version is not often available in the same person. This drive, like the drive for multiple mates in men, has shaped the strategy of women. It has shaped their personalities.

This is also why women tend to have two sexual modes. One for the hot guys and another for the rich/devoted guys. She might show equal sexual willingness for both. With the hot guys, she may consciously feel lust, but it's really her genes who want the genes of the hot guy, and they will allow for a quick exchange to get them. But for the rich/devoted guys, there are strings attached. She will give an ugly rich man the good stuff, but he's gonna have to work for it. He'll need to give her gifts and show signs of love. For the rest of us who fall between hot and rich, she will hesitate. Men will need to convince women of both their quality and their devotion. They will have to work out. They will have to write songs. They will have to dress well, show good manners and social skills, and get good jobs. They will have to talk trash about all those hot and rich guys. And they will have to spend money continuously. And then . . . maybe then, they'll land a woman.

Women can always do better. In 9 out of 10 cases in today's monogamous society, a woman is with a less than exceptional man. That's not science, it's just me looking around. A long time ago in more polygamous societies, she could have been with an exceptional man, but she would have had to share him with other women. This seems to be a viable strategy for humans, though it brings its share of challenges as well. But today most women have to choose among a foray of less than brilliant men. What's more, women *have* to choose! It used to be such that women were "directed" into relationships either by parents or tribal elders. Now, with women choosing and biological clocks ticking, women find themselves in some confusing and often scary territory. A man might like you and want to marry you, but can you do better? Should you hold out? Ask your girlfriends and family and you'll get all kinds of excited information. If you pass on the chance, that might be it for you. Maybe you should just go for it – any deficiencies the young man has can be corrected through proper husband training. Then again, maybe he'll never learn to dress well and chew with his mouth closed. So confusing!

And even if you land the ideal man, there will be fear that you may still have to share him in a sense, since many of these men have mistresses on the side. Think about John F. Kennedy cheating on his wife Jacqueline, bringing girls to the White House for late night romps. As a woman, what do you do in such a situation? The strategies are "grin and bear it" (and hope it doesn't get too public) and try to control it in a game of brinkmanship (I know, that was *his* game!). The couple teeters on the edge of breaking up and staying together. If he leaves, she'll hopefully get the children and a good bit of money to continue without him. If he stays, she could divorce him emotionally and just find a way to live her own life still married to him. She rests assured that she got what she ultimately wanted – his genes and his money. But she did not get him all to herself as she originally expected.

The point is, men don't have to be so choosy when choosing a lifelong mate, since many will expect to get the desired variety on the side anyway. Ditto some women, but to a lesser degree. This is the reason there is a Mrs. and a Ms. but only a Mr. Ever wonder why you can't tell by a title if a man is married? I'll bet now you can guess that men prefer that you don't know. Men also prefer to know whether a woman is taken or not and especially want you to know that **their** wife is off the market, hence the Mrs./Ms. distinction.

I am describing stereotypes – there are many nuances and shades to the above – but stereotypes exist for a reason: They describe the biological strategies behind our behaviors. The male personality is tinged with a more quantity-oriented sexual drive that is less common in women. Are there differences in the sexes in terms of personality? Of course. Differences exist within the sexes, too, and those can be looked at in terms of investment strategy. Are you more of a risk taker or more of a safe-bet kind of person? If you're a risk taker, you might like spicy food or going bungee jumping. These things seem to go together in a person's personality. It then follows that you will be more sexually promiscuous. And both men and women will have their own high-risk/high-return and low-risk/low-return ranges which will overlap somewhere. All men and women are trying to pass on their genes and their personalities — their GPSs — have been built into their brains to help them as individuals to do this as successfully as possible.

And yes, because of the fundamentally different biological realities for men and women, their personalities are different as well. Personality is not some creation of the establishment that tries to oppress women and make soldiers out of men, but rather the inevitable strategy that our sexual equipment and different food-finding skills guides us toward. Men are men and women are women, but we share the same genes. There is no boy gene or girl gene, so the differences in personality are not genetic differences – they are hormonal ones. Nevertheless, nature gave them to us and they are still biological, not culturally created. Men and women both fit into a group and both try to get the best mates within that group, sometimes out of the group, and try to raise their kids as effectively as possible.

We have to constantly remember that a personality is much like a position in a soccer team. Not everyone can be a forward or a fullback, but the point is that everyone *has* a position and they are all important. Each position has its own strategy for succeeding in the game. Carl Jung, the famous psychologist, proposed the existence of a collective unconscious, where people somehow know about all the things there are to know about being a human being. In the light of evolution and the concept of GPS, we can see that really we all just have a built-in understanding of the game we are playing and the different players with us on the field. Soccer has always had the same basic rules, the same goals, and the same positions. Up until recently, the same has been true of humanity. So, perhaps you can imagine an evolutionary adaptation which gives us a subconscious understanding of the rules, the goals, and our fellow players.

Freud famously laid out 3 parts of the personality: the id, ego, and superego. I would like to propose that there are parts of your personality that deal with biological urges and those that deal with cultural urges. In between them, is a third part which deals with a very dynamic landscape. We have biological urges to eat, sleep, fornicate, and avoid pain. We also have cultural urges to be respected, and loved, to teach, and to dominate a situation or group. All in all, not an easy set of urges to assimilate into one being. They will often come in conflict with one another. Worse yet, your fellow humans will want the same things and you'll need to compete for limited resources. Our conscious brain, that "in-between" part of our personality, is the intelligence center set up to help you get it all. And like I said, it's a dynamic landscape, which is made more complex by the people around you. There are hierarchies and many social nuances to be navigated in life. Ultimately, our massive intelligence, and even our consciousness, must have been an adaptation to deal with all the thoughts in **other** people's heads.

There are opportunities to do well in any system and each system has its own set of strategies. Imagine a system that contains a leader and a follower, neither person is better or worse than the other, but one of them dictates the plans, and the other one executes them. Both are essential, and both get the rewards of their success. The follower willingly agrees to follow only on the condition that they share the fruits of their endeavor. The point is, you need both people. You can't have two leaders and you can't have two followers. The same is true of the "teacher/student" and the "good cop/bad cop" scenarios. You need complementary roles. These roles themselves are market opportunities that evolution has filled just like the demand for good coffee resulted in Starbucks. If the leader/follower setup wasn't working, evolution would have killed it off.

In a human system, there are more than two roles. How many roles exactly has been the topic of study of many psychologists.[25] What's important to note in all such systems, is that the personalities described are universal. They are found in all cultures in all time periods. Everyone fits into an archetypal category. And that archetype is simply a strategy for genetic proliferation — a position on the soccer field of evolution.

---

[25] Check out the Enneagram system, which gives us 9, or the Big Five, which gives us 5 scales.

Consciousness, Lying, and Drugs

I want to have a brief talk about a few major parts of our daily lives which touch upon strategy: consciousness, lying, and drugs. So, I've basically said that your personality is just a strategy for genetic proliferation. These "big three" aspects of our lives — consciousness being a key aspect of our psyche, lying being a notable characteristic of the way we communicate with each other, and drugs being a manipulation of our mental experience — need to be talked about in the context of GPS.

Consciousness

One of the most prominent parts of our psyche is conscious thought. We love conscious thought which consistently ranks in the "Coolest Things About Being Human" list. We often think our thought process *is* the same thing as our personality. "The thoughts we think make up who we are, so we can just change our thoughts and become a different person." Wrong! Those thoughts are created by the subconscious, and we don't have a screwdriver long enough to reach that deep into our brains to change it. If personality is strategy, then personality isn't consciousness. Personality is your position on the field and consciousness is your ability to anticipate the actions of your opponents. That's a great little tool, but it doesn't change the fact you're a tough defender or a wily attacker. Follow me?

For all that we don't know about consciousness, we can at least say this: it's an adaptation. It's not a gift from God, it's not a miracle. **Consciousness is an evolutionary adaptation created by natural selection over millions of years to help us proliferate in our environment.** Why is it so unique to humans? Other animals have glimpses of consciousness, but it's different, isn't it? If we compare our consciousness to that of animals, the startling difference is the fact that we can communicate it verbally. Imagine what humans would be like if we couldn't talk, if there were no words. You couldn't think all the cool things you were thinking because there would be no words to describe them. It would just be "feeling." Don't we experience things more clearly and even emotionally through a song or poem that describes a situation in words better than we ever did? Isn't that why we love these wordsmiths — because they, in a real sense, give us that experience through the words that best describe it? So, consciousness is great, but without language, its value takes a real dive. Teach a monkey sign language, and he becomes a different creature to us, but also, no doubt, a different creature to himself. All of a sudden he clearly feels emotions and knows what those emotions are. Surely, the combination of language and consciousness has provided humans with this ethereal life above the pure stimulus-response plain that we see all other animals living on.

So, why do we have the one-two punch of consciousness and language that other animals don't have? Lots of animals have arms and legs, and eyes and ears. Why are we so unique in our level of consciousness? What was so different about our evolutionary landscape that the supreme reward of consciousness was granted? A lion's evolutionary landscape consists of water buffalo, zebra, antelope, etc., so they evolved strong bodies and sharp claws to kill them. What was in our landscape that prompted the development of the most powerful biological mechanism on the planet? In a word: people.

Consciousness, really, is a simulation program. To demonstrate, let me tell you briefly about the exciting world of software development – because really, there are many parallels. Your brain is a computer and your thoughts are software. There are holes in this analogy, but it will do for now. Let's say you work for Amazon.com (because I love Amazon). To simplify, right now there is a big, fat computer sitting on the West Coast of America with a folder called "amazon_live." In this folder are all kinds of files which comprise the code which runs Amazon's website. If they want to change the front page to show a wooden crab hammer for sale in the top right corner of the screen for all visitors from Baltimore, they need to change some of the contents of the "amazon_live" folder. The problem is that a billion-dollar website is a complicated thing, so changing it requires some time. There is also a good chance you won't get the change right on the first try, and a mistake which results in a messed up home page for millions of visitors would end up costing Amazon tons of money. So what does Amazon do? What does any software development team do when making changes to their system? They set up a test environment.

In a test environment, you can do whatever you want. If the change means editing 10 different files, you can start with two files today, do a few more tomorrow, and finish up the rest next week, then have a look at it, fix a few bugs, etc. etc., and none of your customers notice a thing. When it's done, it all works perfectly, then you simply replace the files in the "amazon_live" folder with the files you changed in the "amazon_test" folder, and the new changes are live. Now visitors from Baltimore see crab hammers in the top right corner. If you're lucky enough, everything works perfectly well in the live environment.

This is how software development works, and this is how human development works. **We try stuff out in our heads before we do it in real life.** We don't have visitors to our website, we have people we interact with in real life. And those people are complicated. They act one way in one situation and another way in another situation. They have different moods, different ways of saying the same things, and all kinds of intricacies which have gotten more and more, well . . . intricate since the development of, yes, complex language.

To successfully navigate the swirling waters of human social interaction, we developed a conscious brain. A conscious brain allowed for flexibility that no other adaptation could. Our situation now is very different from our situation 50,000 years ago, but the hardware is basically the same. It is only through our massively flexible brains that we were able to function in both situations.

Make no mistake, despite the incredibly divine thoughts our brains can have, our genes are still in charge. They are the "Board of Directors" who have appointed you to be the CEO and run this company successfully. They know the goal of gene proliferation for humans will be too hard to manage without the flexibility that your conscious brain provides. Your conscious brain is the CEO and your subconscious is the Board. Sure, as CEO, you make bold decisions, give stirring speeches, and connect with customers on a profound level in a way that your Board never could, but the bottom line is still the same. If you don't get a high GPR, you're out, and you can take your Ivy League degree with you.

In our evolving lives, we need to think about what other people are thinking and implicit in that is understanding what we ourselves are thinking. What do things mean? We want to get inside people's heads and anticipate their moves, just like a tennis player wants to know where his opponent is going to hit the ball, how hard, and with what kind of spin. We want to manipulate people and avoid being manipulated ourselves. Consciousness allows for interpretation of subtle and not so subtle bits of information. But why is that important? Because we're trying to get stuff. Money, mates, and . . . more of both.

I was just watching *The Cosby Show* with my kids the other day (yes, before the scandal). Vanessa has a party at their house and she asks her friend, "Did Robert say he likes me, or did he say he *likes* me?" So nuanced. And then there's the "Did he say he likes me *before* Tony told him that I like him or *after* Tony told him." Because there is so much room for interpretation in the things people do and say, we want to be able to play out a few scenarios in our heads before we commit to a conclusion. It is our software test environment. "If he said he likes me before he knew I liked him, then his statement has more value." See how that is basically a few lines of code? "If *user_location* = 'Baltimore' Then *crab_hammers* = True." "If *when_Robert_confessed* < *when_Tony told_him_about_Vanessa* Then *Robert_likes_Vanessa* = True."

All social animals have a certain amount of test software. What we have and they don't is the complex language that gives the scenarios detail and intricacy. But it's the social element of our lives that first puts the "consciousness card" in the deck. All those conversations you constantly have in your head are not just you talking to yourself but you talking to other people. It's no wonder that consciousness is so linked to that feeling of connectedness, even oneness, with other people, with ourselves, with an idea or work of art, and of course, connectedness with the world. **As we're social in nature,**

consciousness is all about connection. If we didn't need to connect, we wouldn't need to be conscious. Spiders aren't conscious and they are doing fine.

And of course, consciousness allows us to dissociate as well. We connect with some things and distance ourselves from other things. It's a constant ballet of forming allegiances and repelling adversaries and negotiating our position among both. At stake is the success of the genes that created us in the form of high GPR. The strategy we use to exist in social groups is our personality – our GPS.

Lying

If consciousness is a tool for improving GPR and not a part of our personality, then what about lying? Some people consider the lies we tell to be another unique trait of humans. "Animals don't lie. How could they? They don't even talk." But people do and we consider their lying to be linked to their personality. Some people are liars. That's what we know them as. Sneaky, conniving, manipulative rats. . . . Is that their personality? Again, no. Lying is just a tool.

Throughout the natural world, whenever there is communication between two entities who do not share the same exact genes, there is lying. If there is competition, there is lying. When people or animals work together, if the goals do not match exactly, as they would in genetic replicas, then there is the opportunity to cheat and get a higher GPR at the expense of the other.

What's the difference between lying and cheating? Good question. They are both deceptions, but one is about information, and one is about action. They are both breaking agreed upon rules. That doesn't mean there is a contract, but there is an understanding in their cooperation, and cheating and lying breaks that understanding.

You see those birds pecking the inside of a crocodile's mouth while he just sits there like he's at the dentist. And you think, "Why doesn't he just eat that bird?!" Chomp! Easiest meal he's ever had. But that would be cheating. They have no contract, but they have an understanding that was biologically programmed millions of years ago: "I'll open my mouth, you can have all the crap between my teeth and I promise not to eat you."

Another animal wants to eat a snake, but the snake rattles his tail to announce that he is a poisonous rattle snake. In truth, he is not a real rattle snake, just a snake with a rattle. He lied.

Lying and cheating both work. Nature loves a good cheater. But the cheated don't like it, and they develop technologies for flushing out a cheater. The best, most relevant example is women. Women have developed a fine-tuned sense to know when men are lying. As we've said, women are the choosy sex, they want a quality man. So, men have developed the strategies of either being high quality or lying about it. In response, women have developed ways to tell the difference. We call it "woman's intuition."

Some people lie all the time and others do it rarely and on a harmless level, like "Oh, you look so beautiful." Some cultures lie more than others. Americans lie all the time ("It was so nice to meet you"). Dutch people, in contrast, are honest to a fault ("That shirt looks stupid on you"[26]).

We all hate liars but nature's cool with lying. What works, works. People seem to be as honest as they can get away with. More on that later. . . . What we need to know about lying for now is that it's not part of our personality, it's just a tool we use. Some personality types will be more prone to use it than others and some people will have learned to lie or to be honest early in their lives as taught by their environment either directly or indirectly. Like consciousness, like everything else, lying is a technology.

Drugs

"Drugs is all around you" as Dave Chappelle said in that crackhead sketch on *Chappelle's Show*[27]. I once saw an interview with Steven Tyler

[26] I know, "get bent, you arrogant sandwich-obsessed, dike-plugging, race of insurance salesmen" . . . I'm so joking, I love the Dutch and their horribly impolite honesty.

[27] http://www.quanology.org/chappelles-show-crackhead-sketch/

who said, "If humans had a button on their shoulder that would give them a quick high they would be pushing it all day long." I thought, "Nonsense—at some point you would get bored with it, no?" Apparently there was a study where they put a button on people's shoulders and when they touched it, it gave them a pleasure shot to the "pleasure center" of their brain. Guess what they ended up doing all day long for days at a time until the experiment had to be stopped since they would have died from lack of eating? Steven must have heard about that experiment.

Our daily behaviors are directed at getting higher GPR. But how do drugs fit into that strategy? We love drugs. From caffeine, to nicotine, to cocaine, to FIFA 15, things that stimulate our pleasure center get a lot of attention. There was obviously a time when none of these things existed and stimulation to the pleasure center was only for things that were good for us, like eating and bonking. We have discovered new substances and activities that are not good for us but also stimulate this part of our brain and now we have a dilemma. Our brain is flawed. **We do many, many things that do not improve our GPR.** We need to go back and recreate the brain to only give us pleasure when we do things that are actually good for us, and not when we do naughty things. Unfortunately the finance department will not justify the huge cost and impact on our current society, so we're going to have to make do with the brain we have.

We all take drugs in some form. Some of us are addicted. It looks like a bug in the system. Wild animals don't become hopeless addicts because they are too caught up in the game of survival and sexual reproduction. Our brain has been wired to get used to any level of the "good stuff" that's available, so that we'll continue to push for more. But it never expected high levels of food or sex, so that internal strategy was safe – like a website that's been poorly built and can only handle a few users at a time. It's fine until it gets thousands a day. Our brains, created for a time when survival meant cocking your spears and crossing your fingers, never anticipated drugs, foods containing lots of sugar, salt and fat, pornography, or slot machines, and therefore has no built-in resistance to addiction.

Even entertainment is a drug. We are the only species that spends time watching moving images and giggling. Concerts, for instance, are just big fests of people on drugs. Music is the drug, stimulating the pleasure circuits of the brain in the same way that narcotics do. I was addicted to Metallica for ages. It was tons of fun. . . . More on art in the next chapter.

Do we get it now? We just spent a damn lifetime reading about "gene this" and "gene that." "The genes are in control." "Everything is investing." "Natural selection is the boss." "I thought we were gonna talk about *me*, man!" Your patience will be rewarded. And we really are moving closer to an in-depth discussion about *you*, but hopefully you already have some new insights about humans that can be applied to you. Before we talk about you as an individual, as promised at the beginning of the book, we need to talk more about humans and the society we've created. You'll understand yourself and your world better for taking the time to do just that. The better you understand yourself and the world you live in, the better you can make good decisions and navigate the streams of your daily drives and the drives of those around you.

You now know that genes are at the root of it all, and that both genes and the cultural technology of thoughts, beliefs and tools all compete and struggle for dominance. How will this new information affect our view on every day human affairs? For me, it changed the way I look at everything. The next chapter will take you on the journey of that new perspective, starting with . . . "Is he gonna say numbers or sex?"

Sex.

4 | The Human Experience

By now, you should have a very different view of human beings. You understand why we're here and what we're doing. You have a new measure to judge how you're doing in life. You understand that you are an animal like all the other creatures on the earth, but through culture you have many more options for your life and way better television reception. Your genes are spread throughout many different people, and you, as an individual, want to see all of your genes passed on. You are here to help yourself but also to help others – not in the direction of wealth or health, but toward those things that increase your genetic proliferation rate. The better jobs, better education, better looks, better group of friends, all have at their foundation the drive to pass on our genes. This is what drives you all day, every day. You also learned that there is a gap in the world we were created for and the world we now live in now that allows us to get the rewards for passing on our genes even though we might not be doing so. Some of these activities are good and some are bad. These things together explain a lot about who we are as individuals and what we are really up to all day long. Given these insights, you should understand yourself better — why you want conflicting things, why you struggle in relationships with other people, and why you just can't seem to get enough.

Now we take a more in-depth look at the world around us. It's important to understand the individual, but we gain some context by understanding the field we're playing on and the rules of the game. What happens when human beings live together? What about the world we created? Being part of our sexual landscape, one of the biggest determinants of our behavior is the fact that we live together. We are so damn powerful because living together has allowed us to work together and create things that we would never have been able to create on our own.

We created this world. Yes, I know you can't stand it sometimes. We have to go to work and to school, listen to the wife nag, watch the husband sit on his ass and do nothing. This is the infamous *system* — the game. You have most certainly said, "To hell with it, I hate it" at some point in your life. You probably meant it, too, and dreamed briefly about living on some desert island where no one tells you what to do, no one messes with you, and the life you live is for you and you alone. Then you calmed down, shut up and kept going. But when the mood hits, you allow yourself to complain about all the bullshit in the world. This chapter is about that bullshit, those human institutions to which we are married: school, work, government, family, etc., and the natural consequences – politics, war, marriage, etc.—that come along with them.

We'll see that just as natural selection has created all that is *you*, so has it created all that is *us*. Everything in our society has been created by natural selection, and the difference between what our society is now compared to what it was then comes from the same evolutionary motor that turned deer into giraffes and this one ape into a modern day human. Remember that mating is the mandatory recipe in *The Slight Differences Cookbook*. In understanding evolution and seeing the control it still has on us, we see that sex has the most profound influence on our society.

Sex & Love

Alright, get ready to blush. We talked a bit about sex, and we know that it's the only way genes get passed on in humans. You don't have to have sex to get your genes passed on, but *someone* does. We're gonna get graphic here. And by that I mean, show lots of charts. Below is a chart showing the number of men who currently have multiple sex partners compared to the number of women in the same boat over the past seven years.

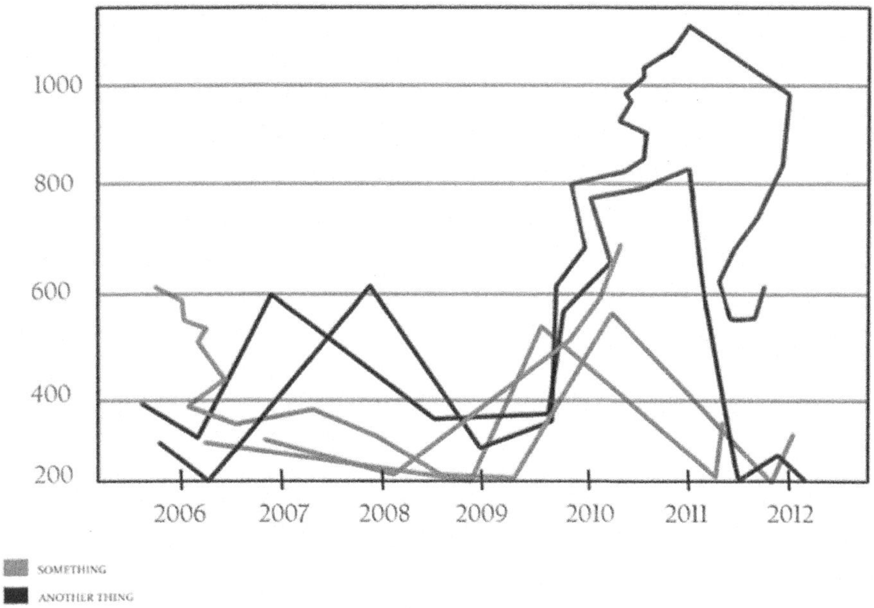

Can you see it?

As we've thoroughly discussed, sex is a big part of life. "Survival of the fittest" is often considered a summary statement for evolution, but as we said, the truth is too complicated for sound bites. We now know that **evolution doesn't care so much about survival, it cares about reproductive success**. It doesn't want you to survive, it wants you to pass on your genes. Survival, up to a point, is just a prerequisite for this. The point of basketball is not the shoes, it's just that you'll need shoes to play well. Unfortunately, the word "fittest" conjures up images of strong, athletic, alpha-dog humans, but we know now that these are not the only types who get their genes passed on. "Fittest," then, should really be "fitting."

So, here's a new slogan for evolution. . . . Say it with me: "*Genetic proliferation for the fitting*." It's got a certain ring to it. There are layers to successful genetic proliferation in humans — strategies on top of strategies — and the bottom two, the foundation, are sex and love, respectively.

Good sex is out of this world. When two people are completely into each other, uninhibited and open, sex is an experience that's like nothing else. It's life-affirming, spiritual. But in your lust-driven haze, you are being manipulated. Nature wants you to do something very, very badly. If you do it, she gives you that incredible feeling that only good sex can deliver. Sex is at the very core of mammalian life, and nature has spent a billion years improving the rewards program that members enjoy. It is very, very important that we like sex. If we didn't, if something inside us told us it wasn't worth taking off our pants, then none of us would be here. After the desire for food, shelter, air, and water comes sex.

As beautiful as it all can be (and as sticky), underneath it all, sex is a numbers game. Like everything else in the living world, it is an investment which needs a return. It costs energy to have sex. Not just the physical act itself, but all the preparation that comes before and the potential offspring afterwards. Because that's why we're doing it. To us, genetic proliferation means babies. We can talk about the importance of sex in terms of physical or emotional health or what it does for intimacy in the relationship, but sex in the Sweet Spot meant babies. Now, we don't have a baby every time we have sex today, mostly because of contraception, but in the Sweet Spot the correlation was much stronger. You weren't necessarily having a baby every time then either. If you were a woman, you may have just been tricking some poor sap into thinking he got you pregnant, thereby making him prone to help out. But the drive to have sex is meant for procreation.

Most of the time we have sex these days we don't consciously want to have a baby. In my experience, I would have liked to skip the sex in conceiving our children. Trying for a baby made the whole experience very unsexy indeed. But to your genes, sex is the road to babies. And that's what your genes want.

So, we want a high GPR and there are a couple of ways to get it. You can be involved in one of them or many. Sex is one way. We know what the reward for sex is: children. But what about the price of sex? What do you have to pay to get sex? Because nothing's free in this world. To have sex, you need something from someone else and you need to pay for it. You may think I'm being crude and making reference to prostitution, and soon I will, but for now I'm just talking about investment of resources – those things we talked about earlier. You have resources, and you need to spend them to have sex.

As I've also mentioned, men and women are fundamentally different when it comes to sex. Men have unlimited sperm and women have precious few eggs, so they invest differently to get sex. That is to say, they pay for sex differently. The fact that sperm is readily available and eggs are hard to come by means that women are choosier than men, and that men need to invest in their status and/or strength to get women. Since men are so competitive they have traditionally dominated status and strength related activities. Men's fierce competition among each other has been the driving force of all that is technologically and artistically great in society. Men are painting ceilings and inventing Youtubes just to win women's favor. There is much investing up front just to get a woman, and men pay in the form of these cultural feats and violent struggles. Naturally strong and attractive men have a leg up on the tech geeks because women want good genes as well as high status. Such men need less investment up front to get a woman. Men with good genes have a sort of "trust fund" to pay for sex in the form of their good genes. Some men have neither status nor strength nor good genes, so they put their energy into lying about it. As men are the cultural pioneers, they are also the best liars.

Women, theoretically, pay less for sex. Since men have so much sperm, most women can find a man who will have sex with them. In reference to pair-bonding in the Sweet Spot, Robert Wright quotes Donald Symons as saying "Every man who can snare a wife does, and virtually every woman is married by the time she's fertile." As humans got more monogamous, men became choosier too. But monogamy is not as deeply rooted in men as it is in women.

In today's society, men have developed a heightened sense of choice. We live in cities and when seeing the thousands of young women walking past, a man might develop the crazy idea that he can bone them all and therefore, becomes even choosier, "playing the field," as it were. That was one of the themes of the movie *Hall Pass* – guys are granted permission to have sex with other women, but they soon realize they are still bound to the laws of human nature. Women, too, see "other women" more than ever, and the initially soft drive to look pretty now becomes an urgent one in the face of all the new competition. They now need to invest more to get a good mate.

Why do men even care that women look good? Because attractiveness denotes good genes. But unlike female attraction which tends to favor strong, physically fit men, male attraction tends to favor young women. This is because young women have more potential children to give and have a better chance of bearing healthy children. So, all other qualities being equal, we expect a man to choose a younger woman over an older woman. Men don't need to be as young as women to have children, they only need to be able to afford them, which is why women fall for wealthy men, and wealth often comes at a later age. And yes, women also lie about their age and the quality of their genes by wearing make-up and high heels and spending hours at the hair salon. Again, because of their endless sperm, men don't mind being fooled all that much.

So, we've all invested. **But women have invested more**. It takes more effort to create an egg than a batch of sperm, and what's more, eggs are more scarce (12 a year compared to millions every day). Investing more means that women will also be more driven to provide their potential babies with a quality upbringing. Whereas men are genetically paranoid about women cuckolding them, women are paranoid about men abandoning them after the baby is born. To solve these problems, they have forced men to invest more in the game by proving their worthiness up front. "Take me out on dates, buy me flowers, say all kinds of romantic things which are really just promises of future commitment, and then maybe I'll be your girlfriend." The cliché dynamic between men and women is natural and biologically rooted. Most men are working hard to win women's favor and most women are doing their best to be the object of affection, all the while demanding proof of worthiness up-front. In a perfect world, both camps end up in bed with the same amount of investment. But here's the rub, we each want a better deal. We all want a better mate than we are worth. Not all of us get one, not all of us put much effort into getting one, but we all want it. Every guy wants a girl who's out of his league, and every girl wants a white knight. **So much of humanity's progress is a result of this desire.**

But before we look at the fruits of this phenomenon, we need to look at what happens after sex. *Ain't talking 'bout love*! No, wait . . . I *am* talking about love.

"Wait, why is love *after* sex? Don't you love people before sex?" Yes, but love is the emotion that says, "I want to continue on with you *after* sex." For men it's "I want to help raise that baby once you have it." For women it's "I want you to be the father who takes care of our family." This emotion may appear before sex, but it's looking into the future — a future of investing and partnership. Of course, we're talking about intimate love here, which is different from parental love. **I want you to see love for what it really is: an "emotional contract."** If two businesses work together for a common goal, they draw up a contract and both sign it. When two people want to partner to pass on their genes, they also have a contract. It's usually verbal, and it usually involves the word "forever." In recent centuries it has been formalized in the form of wedding vows — "honor and cherish," "till death do us part," and so on.

Even lust is a sort of contract. That contract might be for to a short-lived romp in the hay. Here, the focus is on the short term — simply conceiving the baby and not looking too far into the future. But, at the very least, nature would like you to be in lust so that you'll have babies. It understands that two people who are equally excited about the prospect of conceiving and raising a child is less common than two horny members of the opposite sex. If humans had to wait for groups of two people who were totally in love to raise babies, we would not have come nearly as far as we have. In other words, lust will do just fine when real love isn't around. The group you live in will jump in to impose rules and social pressures regarding what you do with the baby you might have. Those social norms will be a handy substitute for the missing "love contract."

Now, love is not a contract in the traditional sense, and yes, it can be broken. People fall in and out of love all the time, but the point is that it signals a long-term commitment. I'm not talking about anything "sacred" or advocating the idea that love is above biological urges. It's just a strategy — one that has helped us raise more quality children as opposed to children with limited parental support. **That's what love is: an agreement on parental support.** If I find a woman I like and who I deem worthy to further invest in, I *love* her. My judgment of her on that level is purely subconscious. Consciously, I am dumbfounded and completely infatuated. I think about nothing else. Yes, I want to shag her, but the visions in my head are of vacations, romantic dinners at home, away from the rest of the world. We have sex, but we're also cuddling. We're pushing strollers, carrying the kids on our shoulders and getting old together. I'm buying her clothes she doesn't need and jewelry she doesn't expect. I'm making a great impression on her friends and family. Get the picture? Sure you do. But my subconscious is laughing and telling his friends about what a sucker I am. My subconscious has hypnotized me and I fell right into his trap. It saw a girl who was pretty (genetically sound) and noticed a few hints which suggested she would be a devoted and caring mother (think warm smile and classy clothes). It also noticed her showing a slight interest in me. The hints were subtle, but there — she held eye-contact, she smiled during conversation. So, my subconscious told me I was in love, and this emotion made me move heaven and earth to make her my wife.

This actually happened! I was only 20. She was 22. A tall blond Dutch girl of proper upbringing and with an open desire to be a mother someday. She was beautiful and still is. She agreed to go out with me and that was that. Almost 18 years later the contract is still in place. If I didn't love her, if she didn't love me, we would not be raising three kids together. The sex is over . . . eh hem, I should say, "The conception of those three children is over." So, it's the intimate love that sees to it that we keep working together.

For now, we just need to know that genetic proliferation means that somewhere two people must be having babies. This means we need a healthy dose of lust to get to that point. From there we need to invest to make ourselves attractive to a potential mate. The ever-present horniness and "happy feeling" in our pants pushes us out of our clothes and into bed. If love is present, the partnership has a good chance of continuing into the future when all cylinders need to be firing to raise kids. This is how nature built us – different from many animals, similar to some. We are a slightly polyamorous animal who lives in groups, competes for mates, and has a high level of parental involvement. These sexual factors are the real cause of our high tech society, and the drive to get more out of the deal ensures that we continue to innovate and will live in the soap opera that we do for the foreseeable future.

Men versus Women

> ***"All I can say is thank you and thank God I found you. You've given us three beautiful boys and we have a wonderful life together. But I do have to say sometimes you get a little lippy. OK? You've got a big mouth and . . . you like to run it."***
> – Will Ferrell's acceptance speech at the Mark Twain awards.

Let's start with the most basic group-unit of humans – men and women as intimate partners.

> ***"If you don't like fuckin' somebody, and you don't like eatin' with them, y'all don't need to be together."***
> — Chris Rock

This chapter is called "Men *versus* Women" for a reason. **Men and women compete as much as men and men compete**. We all want a good deal and a good return on our investment. Ever wonder why it's possible to start hating the person you loved yesterday? Or even the other way around? **Sexually, men and women are in a zero-sum game**. We each want to get 100% of our genes passed on but we can't both get 100%, so we compete for the largest share. Selective pressure has ensured that they each get about 50% in the baby. Why? Because those genes which did not get in at least 50% of the new babies died out. So, successful copulation squares us up at even. But before that, much has taken place.

Men want to dominate women, and women, yes they do, want to dominate men. A life of surplus, which is what our species has at the moment, leads to men dominating women. Men hold power over women and other men in the form of wealth. Women hold power over men in the form of sex. Isn't it all so romantic?

Sexual selection also has ideal situations and extremes which are rarely seen but which demonstrate the ideals and tendencies of male/female pair-bonding. Our subconscious pushes us toward these behaviors in moments that allow for them. When the door is open, we try to get more. These ideals and tendencies are not always easy to hear about, so be forewarned. Here we go. . . .

Ideally, a man is rich and has a wife, plus a few mistresses on the side. Do you think men want to be rich and have a wonderful wife, but also enjoy a sexual liaison here and there? Not all of them do, but there is a slight pull in men toward this goal. I expect you recognize this.

Of course, they aren't as eager to let their wives enjoy the same sexual indiscretions. We have seen why this is. A man is jealous of any attention his woman gives to other men, as it could result in him raising someone else's baby — a genetic catastrophe. As much as your man thinks you're over-reacting when *you* get mad at *his* flirtatious behavior, he hates it when you enjoy the charms of another man. If he does not get annoyed with flirting, one could say, "He's very secure about himself and our relationship." But another interpretation could be that he is not giving much to his current relationship with you and therefore isn't spilling too many resources if it turns out you actually are pregnant with someone else's baby. Maybe a guy who is not the jealous type is also not the devoted, giving type, if you see what I'm saying. . . . Maybe the jealous guy is giving more and therefore has more reason to be jealous.

Male sexual jealousy is theoretically stronger than female jealousy, because the impact of a cheating woman is greater than the impact of a cheating man. If your man cheats, it's possible that all he's done is spent a bit of time and sperm on another woman. You won't miss the sperm, I promise. And even if she got pregnant, it's possible that neither of you would find out about it, and therefore, would not be burdened by it. But if a woman cheats, it's possible that she could get pregnant, and her current spouse could find out and decide to leave her, or she could lie and tell him the baby is his. In both scenarios he's killing his GPR. So, both sexes are jealous when it comes to sexual infidelity, but male jealousy goes deeper.

And now a woman's ideal. **Ideally a woman has a wealthy man all to herself**. Know any women who would like their man to be rich? Again, not every woman has a rich man, but a rich man would be their preference. **Ideally that man has great genes**, so she need not look elsewhere for better. If he does not have great genes she may be slightly inclined to seek out trysts with other, better looking men. Women want to dominate men as much as men want to dominate women. A woman has an advantage in the beginning of a relationship when she is enjoying the fact that she can be choosy. They all want to stick it to her and she knows it. It's her time to call the shots. The men with the greatest potential (yes, GPP) and the most overt desire to dedicate it to her will have the best chance of having her. She expects men to put on displays of power and worthiness. When she chooses one and "gets" him, she will expect total devotion as promised. If he doesn't deliver, there will be jealousy. Yes, women will also be sexually jealous as we just saw, but their most potent version of jealousy is emotional jealousy over their man having *feelings* for another woman. Women fear spending all this energy to have a baby, and then being left in the cold to raise it.

So, when a man spends too much time in the pub or spends time "talking intimately" with other women, jealousy will not be far behind. For her, her husband's emotional devotion to another woman and spending lots of money in bars is taking away from his ability to provide support for the baby. Unacceptable.

Men and women are by nature at odds with each other, and in light of all this subconscious manipulation, the fact that they have harmonious relations is a strong confirmation that natural selection can and does lead to happy moments in life, aside from all the fulfillment of GPR.

The Phases of Love

I hope you see the struggle that men and women are involved in. Now, let's have another look at some of the applications of sexual strategy on our personalities by breaking down men and women into *phases*. We all understand that there are phases in relationships which correspond to the work at hand. The first phase is courtship.

You like someone. Maybe you *like* them, like them. Maybe they like you back. It starts with attraction. First, generalized to a gender ("I like girls"), then focused on a specific person ("I like *that* girl"). You want them. Maybe you want to be with them. Marry them. Sleep with them. Maybe you just want them to want you. Maybe you're not sure what you want. But you know that *they* have to give it to you. . . .

And that's where the hate comes in. You can ask for affection lovingly, but if it's not given, you can quickly switch to "Give it to me, you bitch/asshole." Either way, it's for them to give and you to get. Sure, we all think of love as something selfless, but through the lens of evolution, we can see that it's really "expecting." **Love is not selfless, it's quite the opposite**. A person who loves and wants nothing back is a perversion. It's a fairy tale that many of us have told ourselves. And it's a lie our subconscious doesn't mind that we believe. I've been there. I've said nonsense to myself like, "I just want her to be happy. It doesn't matter who she's with." But that's horse puckey. I want her with *me*. The beauty of my self-deception is that she is more apt to believe that I care about her more than myself. Not only my words, but my entire being conveys total devotion. Just what she's looking for in a guy.

"No one loves you as much as I love you."
– the gist of like a million songs

Ultimately, women want selfless devotion from a guy. You may be ugly with no career prospects, but what you have you will give to her. "Other guys have more than me but they are selfish." This is the sales pitch of the hopelessly-in-love. The more they themselves believe it, the more convincing it will be to the customer. Sorry to burst your bubble, ladies. It's not going to be a free ride for long.

Women are known to fall in love more frequently and less deeply than men. Is it any wonder? For a man, settling on a woman is a big deal. We're more wired for multiple women, so our genes tell us not to fall in love, real love, too easily. When we do, it opens the floodgates. Women, however, are more wired to find a good man and grab him. And a good man is one who spends a good bit of effort showing how much he's going to give to a woman. This puts women less in the "lover" category and more in the "lovee" category. Women don't typically go so head-over-heels for a man.

They can get really excited by a guy but they don't show it readily. They might be up for a quickie with a good looking guy in the broom closet at times, but deep devotion is not their thing, since it's not the grand displays of love that will land them the man of their dreams, but rather coyness. Women are better at behaving like rabbits in order to get men to pursue them. Provide little hints that you're there and catchable, and the fox will instinctively give chase. A little eye contact here and there, a touch on the shoulder, a casual aside that you are not happy with your current situation, a little flick of the hair – all while dressed well — should be enough to set him off. From there, the hard-to-get strategy tells the man that you are to be classified as "long-term relationship," not "broom closet romp." And, yes, that classification exists in every man's mind.

Ladies, if you want a romp in the broom closet, the object of your affection would most probably agree to take part, but with it comes a reputation as a "short-term gal." But, then, no one needs to know about it. And this is how some women make their living, and many others fantasize about — secret flings with the genetic elite and the coy-ploy with men of high status. As we have explained earlier, ideally women get good genes from good looking men and good resources from high status males. So, just as men categorize women as long-term and short-term, women do the same. Some men are for boning and others are for carrying Prada bags and opening car doors. Men know this and resent women for it. Men on the status side call women "whores" when their sexual focus goes astray. Actually, men want women to be loose, but only with them.

And so goes the struggle between men and women in courtship. He wants sex with little effort, but barring that will put in the work for the women he deems worthy of his investment. He wants to maximize GPR, so he wants pretty (genetically sound) women who are of child-bearing age (13-40). If he characterizes her as the long-term type and sees that she may be interested, he will be more inclined to buy her all kinds of nonsense and send her poems about her eyes.

She wants sex, too, but prefers one good shag with a George Clooney than a number of shags with Paul Rudd-types (yo, I like Paul Rudd, too).

But if Paul puts in the effort and shows his devotion, she will consider him as a long-term partner in her quest for high GPR. He'll have to dance, wear nice sweaters, talk charmingly to her friends, seem serious about work, and casually mention how he's not really interested in other women. And doesn't that just sum up Paul Rudd!? If she's convinced, she may settle for him. If not totally convinced, she may just *seem* to settle for him, gladly accepting his generous resources, but look elsewhere for some better genes, or barring that, be disappointed her whole life.

These are the ideal situations for men and women genetically. Some of us have a hard time recognizing ourselves in this, but we may know a few friends who fit the bill. Then again, maybe that friend is you. Life is largely about compromising and most of us have hopefully found someone we're happy with and only feel brief moments of dissatisfaction. Long-term, faithful relationships are the norm in a stable community.

So, maybe you're not trying to score a new woman and you're not scheming to trick your man into raising someone else's baby, but there may be moments when you're attracted to members of the opposite sex and other moments when you are bored with your current partner. In some people these moments are brief and insignificant. In others they are pervasive and intense. If you're young you may be wondering who you'll end up with and what your love life will look like when you're older. Most of us hold a conscious ideal of a long-term partner who we feel blissfully connected to. In this mental picture there is no temptation and no dissatisfaction – not on the wedding day, and not 20 years into it. But we know this is rare. And your genes are happy to keep it this way. **It is your genes who push you to look at members of the opposite sex despite being in a happy monogamous relationship, and who ask your partner to be more attentive, harder working, more loving, and more amicable than they naturally are.**

It's clear that relationships are a struggle – always have been. This is natural and not some fault of modern society. But for now, let's imagine, two people happily committed to each other. Let there be sex! Go on, picture it. Clothes coming off. Kissing, touching. Some noises. Intense eye contact. *Take my Breath Away* playing in the background. A passionate kiss between thrusting. Words of love. Switch positions. More words of love. Keep it going . . . and . . . done. . . . whew. . . . That was . . . incredible.

But once it's over, the real work starts. Early in the relationship your woman loves to get surprises, but later she is annoyed at you buying all the wrong things and taking her to all the wrong places. Why? Because it's a waste, and there is little room for waste in a family economy. Yes, it was a hell of a job finding a mate to have kids with, but now there is a massive shit storm on the way. When that little bambino pops out, your life will never be the same. The man and the woman did this all in the name of increased GPR, but it will all amount to nothing if that little baby doesn't grow up and become a high GPP mom or dad him or herself. If you have ever raised a child, you know damn well that there's no way it's worth it. Unless you add love — that secret ingredient in the pesto sauce. It's the pixie dust that gets your brain to think it's a good thing to wake up at six am every morning, including weekends, and spend countless hours cooking dinner, cleaning up after dinner, cleaning clothes, keeping them out of danger, educating, vaccinating, interviewing babysitters, moving to better neighborhoods, calling teachers, watching countless plates of freshly prepared food succumb to gravity at the swipe of a little arm, etc., etc. Yes, this is parental love at work, and we'll get into that in the next chapter. But before there are kids, there is often (not always) intimate love.

As we said, intimate love is that contract which expresses your willingness to engage in a long-term partnership with a person. Once the baby is there, what often happens is that the love between partners diminishes. Not right away, of course. In those initial months babies need all the help they can get just to survive, so they prefer to have both parents around. Nature has seen to it that the parents get a boost in the love department when raising a small child. I've got a nine-month old baby and I'm sky-high on oxytocin right now. For parents to go their separate ways so early in the game would be a certain waste of their efforts up until this point. Once the baby is past a certain age where survival is almost certain, that's when the dads typically start to look around for new options while the mom tries to squeeze more out of the deal.

I'm not saying that men consciously think "*my child is old enough to survive, so I can now go find another, younger woman to impregnate.*" And women aren't consciously thinking, "*I've gotten what I want out of him, so I don't have to be so nice anymore.*" I'm saying their genes have trained them to maximize GPR, and an effective strategy would be to do things like this. So, it is probable that a man has a subconscious inclination to get restless in his relationship once the kids are stable, and small irritations or environmental expectations may make him consider divorce. Statistically, most male initiated divorces do happen when a child is young, but not too young. There is the infamous seven-year itch, when couples start to reconsider their contract. Seven years into a relationship 50,000 years ago would have resulted in one or more kids of ample age, which would mean "My work is done here with this person" for men and "He better be just getting started" for women. Again, it's about maximizing GPR and we have to think, "What would the genes want?", "What's a good strategy?" and from there try to link our typical behavior patterns with the strategies we think the genes would follow.

Raising kids is hard work, so it follows naturally that the romantic love felt in the courtship period wanes. It costs a lot of energy to be in love, and men and women will need that energy to raise the kids that this love brought them. Sure, there are exceptions — my wife and I are still in love, but no, it's not quite the same as when we were in our early 20s — though we still have our moments (before she reads this and reminds me of several occasions). Usually those moments are when the kids are away. Once you start working on your lives together there will be expectations and the love you experience at that point depends greatly on how well those expectations are met.

You were in love and you signed a contract. Not literally, but you had expectations about what the future would be like with your lover. The future is now, and you are constantly grading that person on his/her fulfillment of those expectations. I expected my wife to be a good mother and loving wife. She's doing a great job. She expected me to be a faithful, loving husband and father, and a well above average earner for the family. . . . The first few are working out great, but, hey, nobody's perfect. The rub, the fights, will often be about those things we expect but are not getting. "How are you fulfilling my goal of higher GPR?" This is what you both are constantly asking, and all your activities are expected to be directed to that goal. I go to work every day, I help clean up, I help take care of the kids, and I don't spend any money/energy on other women. My wife takes care of the kids, cooks, cleans, and assures me those three little buggers are mine. Mission accomplished for both of us.

If it all sounds very calculated and cold, it is – it's just unconscious. We have a warm relationship, as most couples do, but understand what's beneath it. It's a joint venture and both parties have obligations. If you think that's not true, that love is more transcendent than I am giving it credit for, try slacking off for a while and see what kind of a response you get from your partner. Try "Honey, I think I'm gonna take a year off work, travel around a bit." Ladies, try "I'm sick of wearing make-up and sexy underwear. Can you cook dinner this month? My parents are coming over to stay for a few days . . . and then a few more days." Deal breaker!

Look, men and women in a loving long-term relationship are a beautiful, though incredibly boring, thing to watch. Such partnerships have propelled mankind to great heights. It has allowed us to limit the amount of time and energy we spend on romance and sex and get down to the business of working and raising high-quality progeny. It's a strategy for genetic proliferation – one for which nature provides frequent emotional rewards. That is to say, "Kids are so cute," 'cause why else would you do it?

Theoretically, a long-term monogamous partnership is not the ideal situation for a man or a woman. It's a compromise which allows most members of society to have a slightly less than ideal situation – like a highway with a speed limit. You could have a left lane with no speed limit and a right lane with a speed limit of 30, but that would be a bit unfair and the ones in the right will eventually try their luck jumping into the left lane, possibly killing a few people. If you set a speed limit to 60, then on average, most people would be going faster than in the other scenario. No one set of genes is doing exceptionally well, but across the species, they are doing the best they can. Being such a compromise, marriage is naturally prone to fits of frustration.

So, men and women cooperate and compete, each couple with their own bias toward one or the other. **You each want high GPR and there lies potential in the other person to get it**. Like any other cooperative effort, you can work together harmoniously or more maliciously. In every relationship there will be elements of both. Neither of you will be doing this consciously, and most of you (well, not you, you're reading this book) will have no idea what the underlying reason is behind your struggle. Even when you become conscious of it, the drive is too deep to be turned off. Realize that struggle and competition are a part of relationships and through education you can at least get a bit of perspective, peace, and understanding.

A Word about Homosexuals

I've been talking about heterosexual activity as if that's all there is, which we all know is not the case. Of course, nature has also focused on heterosexual activity since that's what's needed to proliferate genes. With this must-have requirement in mind, homosexuality has been a mystery forever. How do homosexuals fit into this puzzle we're putting together in which males and females are attracted to each other, mate, and eventually have and raise babies? Are we homosexual by nature or is it just a phenomenon of modern times?

Modern research all the way down to archeological findings confirms that homosexuality is indeed a creation of biological evolution. It is a natural behavior. It is not some social phenomenon or creation of modern civilization and certainly not the grand vision of sexual-political miscreants looking to destroy the pillars of faith-based society and create a free-spirited cult where everybody bangs everybody else, as many homophobes would lead us to believe. It's always been a part of humanity and looks like it always will be. It was there when we were grunting and hunting on the plains of Africa. Boys were humping other boys and girls were rubbing other girls.

In studying other animals, we see that homosexual behavior is also not uniquely human. Evolutionary biologist Loren Sackett at the University of Colorado counts 217 species of mammals, including all the great apes (of which we are one) that engage in homosexual activity.[28] In humans, homosexual activity has been documented throughout our history and there are even 10,000-year-old rock paintings depicting homosexual sex[29]. The Greeks did it, the Celts did it, and on and on, and even cave men did it. So, you don't have to like it (and I'll try to explain why you don't like it in a sec), but you really do need to accept it as part of the deal.

Now here's where it gets tricky. I want to place the homosexual puzzle piece into this model and see how it fits. Some pieces fit well in this model and some a bit more uncomfortably. But the homosexual puzzle piece itself is not very clear so I'm at a disadvantage before I even try to place it. The reason the homosexual puzzle piece is so unclear is the distinction between homosexual behavior and homosexual identity. It looks like throughout much of history, homosexuality was something you did, more than something you were. In the modern world, we think about people being gay and then having a gay lifestyle. Homosexuals like Elton John come out as gay and that's it – it's a thing that changes their sexual future. No more sex with women, no wife and kids. But it looks like at

---

[28] Article The Evolution of Homosexuality: http://www.colorado.edu/eeb/sackett/LorenSackett/Evolution_Blog/Entradas/2013/5/8_The_Evolution_of_Homosexuality.html

[29] Nash G (2001) The subversive male: Homosexual and bestial images on European mesolithic rock art In: , editor editors. Indecent exposure: Sexuality, society and the archaeological record. Glasgow: Cruithne Press; pp. 43–55.

many moments in our past, heterosexuals acted gay but maintained a predominantly heterosexual lifestyle in parallel. Many heterosexual men in ancient Greece had regular sexual relations with other men while they led an otherwise normal straight life. That's a very chewed up puzzle piece.

Thankfully, we can often gain clarity into human motives by looking for similar behavior patterns in other animals. If we do it, other animals probably do it too, and since they don't make up all kinds of stories explaining their own behavior we can study it unfettered and then see how it relates to us. We know that many animals engage in homosexual activity – males and females. But would an animal ever identity as homosexual? Are there apes that are strictly homosexual? A tough but important question to answer. And I won't do it here. It would be a whole separate book. But, what we can say conclusively are a few things:

1. Homosexuality is a natural phenomenon
2. Homosexual activity has been practiced by heterosexuals at various times and places in human history
3. The inclination to homosexual behavior is a spectrum – stronger in some, absent in others
4. Humans are the only species we know of that have exclusively homosexual individuals

These are the edges of our homosexual puzzle piece and now we see how it fits in our model.

At first glance, it doesn't fit well at all. When we say that the supreme drive in all animals is to pass on their genes and when we as mammals do that mainly by procreation, then homosexuality is quite the curveball. But through the gene-centered view of evolution, we see that procreation is just one strategy in our quest for high GPR. So, what about the others? Does homosexuality fit into the other strategies that might justify its cohabitation alongside heterosexuality? And what causes homosexuality in the first place?

As we're distinguishing between homosexual behavior and homosexual identity, we need to tackle them separately with regards to these questions. Homosexual behavior can be as superficial as something people do to get off. An example would be sodomy in male prisons. But there are even young male dolphins who engage in homosexual activity when finding a female mate is too difficult[30]. Those involved are not really homosexuals, they are just engaging in homosexual behavior for a lack of other options. But of course, we know that's just one spot on the homosexual spectrum. Down further are people who are occasionally attracted to members of the same sex but not enough so to break cultural norms. Some people like both men and women and have intimate relations with both. And then further down are people who are consistently attracted to members of the same sex and less attracted to members of the opposite sex. They are homosexual. Why is that?

Well, once we see how homosexuality could lead to higher GPR in some individuals, or at least not lower it, we realize that pin-pointing the cause of homosexuality becomes less interesting. It's like who cares what causes humans to speak when we just concluded that speaking is beneficial? I will however reaffirm the genes' role determining sexual orientation. Somewhere in humans and other animals is a genetic propensity to engage in homosexual behavior. This means we can't get rid of it through social engineering. Environmental factors may encourage or discourage it, but it's inside us and will be inside our children and on and on for many generations to come. Like most of our drives there is a genetic foundation that is manipulated by factors before birth and then after birth for a good five years. From then on, you pretty much do and like a set of things a certain way for the rest of your life.

But of course, homosexuality is a uniquely mystifying thing – perhaps because it involves sex, that fundamental act that our existence hinges upon, yet seems to miss the point entirely. But when we create a bigger picture we see that it seems to fit well enough to justify its place in this massive puzzle. A few scientific conclusions from the past decades confirm this fit.

---

[30] Bagemihl, Bruce (1999). Biological Exuberance: Animal Homosexuality and Natural Diversity. St. Martin's Press. ISBN 978-0312253776

The first bit of evidence is a genetic numbers game: studies have found that maternal female relatives of gay men have more children than maternal female relatives of straight men. To simplify, imagine a brother and sister pair. Scientists see that when the brother is gay, the sister has more babies.[31] Remember that the brother and sister share a good number of the same genes. What scientists are concluding is that whatever genetic elements there are involved in sexually preferring men, both the girl and the boy have them. The boy who has them expresses this preference by being gay. The girl expresses them by being more sexually driven than the average girl. And through her heightened libido, she has more children than the average girl. This is a form of *sexually antagonistic selection*[32], where certain genes are at an advantage in one gender and at a disadvantage in the other.

We said already that humans don't necessarily increase their GPR by having more children since those children need care to successfully reach sexual maturity themselves. But what if they had an uncle who could help with the child rearing - one who didn't have children of his own? This has been called the *Gay Uncle Theory.* The sister would have lots of kids through increased sexual activity and bypass the child-rearing limitation we just talked about in the Quality vs Quantity section by having her brother help raise them. And this is the next trend that science is seeing – that gay relatives take better care of their nieces and nephews than straight relatives. This is nothing more than tried and true kin selection at work. In this set up we see brother and sister pairs as economic units working together on raising their collective GPR. They share a common genetic trait of sexual attraction to men, and, in their collective efforts, that attraction pays off in higher GPR.

---

[31] The Journal of Sexual Medicine http://onlinelibrary.wiley.com/doi/10.1111/j.1743-6109.2012.02785.x/epdf

[32] Camperio Ciani A, Cermelli P, Zanzotto G (2008) Sexually Antagonistic Selection in Human Male Homosexuality. PLoS ONE 3(6): e2282

Now, if you think about this in your own life, you may have a gay uncle who is quite negligent. "This disproves the theory", you suggest. But again, this is all based on statistical models of large groups and while your Uncle Jack is never there and if he is is usually drunk, statistically scientists are seeing the opposite – that gay uncles are more caring and involved in the lives of their nieces and nephews. This too could have been the case in the Sweet Spot when we were our more natural selves and we were all living together in the same tribe.

And what about aunts? Good of you to notice I've only been talking about uncles. It seems the scientific community is not throwing gay men and women into one batch. If they are indeed correct that there is a large genetic component to homosexuality, then obviously the drive for homosexual behavior is present in both men and women in the same fashion. But the expression of homosexual behavior in men and women could certainly be the result of different strategies. Personally, I find the strategy of the gay uncle applicable to women. Why, as with the gay uncle strategy, would there not be a gene that both pushes *men* to be promiscuous (and thereby have more babies) while simultaneously pushing female relatives to also be attracted to women *and* be more caring of their nieces and nephews, and perhaps even brothers and sisters if it is the father that is promiscuous rather than the brother – a *Gay Aunt Theory*? But I can find no research that confirms this flip side of the coin. And the scientific community in general seems to be lacking a good story behind the roots of female homosexuality.

Maybe the reason why this finding only seems to apply to gay men has to do with that basic, fundamental setup of men and women: that men compete more ferociously for women than women do for men. Remembering that our behavior now is based on our landscape back then, we take again from Donald Symons who estimated that every woman in the Sweet Spot was married in her teens. We can apply this to homosexual women too. Imagine a woman saying to her tribe "Sorry, I'm not into guys". How would they respond? Would they be respectful? Or would the men who might be dead in next month's raid, who fight so hard already to get women simply shrug and throw her over their shoulder anyway? She may not be into guys, but they are certainly into her. And her being more comfortable with women would not have stopped her from having a few children of her own. In this respect, we can imagine that the *Gay Uncle Theory* is more pertinent to men since any man who decided not to marry in the Sweet Spot would not be such a horrible thing for the other competing males. This is true today when straight men joke about the knowledge of a gay man by responding, "Less competition for me".

And maybe homosexual men are happy the competing males have taken them off the radar, which brings us to our last theory - the *Sneaky Fucker Theory.* That's really what it's called. It was first coined by biologist John Maynard Smith who devised the theory after seeing how subordinate male deer were copulating with females while the alpha males were battling each other. This sneaky behavior has been witnessed in other animals too. There is even a cuttlefish that changes its color to look like a female to be able to sneak past a dominant male and copulate with females. No, we're not cuttlefish, but certain strategies, like camouflage or posturing to look bigger, are seen across the animal kingdom and we use many of the strategies seen more typically in other animals. This one is called *sexual mimicry.* Imagine in humans that acting less manly or less feminine took you off the radar of competitors and made your intentions less direct to both competitors and sexual targets. So, if you act like a homosexual man, then the other men will not be as typically competitive towards you and women will drop their guard around you, since neither thinks you're secretly trying to have sex. But then, one lonely evening while her husband is out for a week-long hunt, you and a lady get together and go for it. She gets pregnant and you get your genes passed. And no one is the wiser. This will not happen all the time, but it will happen, and it will keep the genes that support this behavior in the gene pool.

The same for homosexual women. No need to be involved in all the verbal competition happening behind everyone's back. Take yourself out of that ugly game by calling yourself a lesbian and probably at some point a man may copulate with you anyway. You'll be a single mom, but that's OK – your girlfriends will help you out with Junior here and there when the workload gets to be too much. In highly competitive societies, we should see a larger percentage of the population looking to such strategies as ways to avoid the massive pile-up of people competing via more direct strategies. Why walk into the bar and compete with 100 men for the women there when you can just get together with one of the ladies later for lunch? Totally innocent to the outside world and to their conscious minds too. The presence or lack of heavy competition should underline the idea that although there are genetic components to homosexuality there are also environmental cues that push individuals up or down the spectrum – all still with the goal of high GPR in mind.

Competition is hard, so we look for ways to avoid it but still try to get the same results. If society tolerates, and it is doing so more and more, we should see people experimenting with different sexual strategies. And we do. We see more transgenders, gender fluidity (people who seem to go in and out of genders), polyamory, etc. All bucking traditional trends in order not to compete with the rest of the world that follows those trends. But in the end, they all still want that hard-wired almighty goal of higher GPR. Many homosexuals in the modern world will lament the fact that they don't have children. We see that historically their homosexuality would not have prevented them from having children, or at least helping raise them. Why? Because this theory says that homosexuality is partly a guise and that gay people *would* be having children in the Sweet Spot because that's still what the underlying drive is. Some of them still manage it today by initially being in a heterosexual relationship, having kids, and then coming out as gay. But these are painful situations for all involved (even though GPR was increased), and I can only see them as a creation of the modern heterosexual/homosexual dichotomy. In that sense, the modern world pushed homosexuals into a more black and white set up and robbed them of that layer of life. They may be gay, but they still want high GPR and kids are the best, most fulfilling way to get it.

So, it's here and it's natural. In my opinion, it also fits quite nicely into the model drawn in this book. Then why has homosexuality been such a controversial topic for as long as we can remember? It seems a certain degree of aversion to homosexuality is also natural. Parents naturally don't want their kids to be gay, societies don't want their people to be gay, and heterosexuals (especially men) don't want other people thinking they are gay. I'm certainly not condoning homophobia or any discrimination, but if we're all for gay rights, and to be clear, I am, then we at least should, as a matter of effectiveness, ask the question why they don't have them in the first place. Why would the vast majority of modern human societies not treat homosexuals like anyone else? As with most trends there are several reasons for its existence. And as with many things that evolution tries to explain, I don't have the definitive answers or conclusive evidence – just theories and support from other conclusive ideas.

If we want to be real about this, and we do, then any talk about homophobia should first take a look at discrimination in general. The fact is, we like people who are like us and dislike people who are not like us. This is simply genes taking care of their own. We can visually look at people and see that they are like us, but what about the inside? You have to sit down and have a god-damn conversation with people until you find out they love *The Bachelor* before you decide to hate them. The unique thing about homosexuals is the very fact that they define themselves as such. We see that other animals engage in homosexual behavior but that none we know of do so exclusively. Why do humans? You can bet the reason is complicated. A glaring difference with other animals would be the sheer fact that we have the self-awareness and articulation to do so in the first place. We, unlike other animals, are able to consciously look at ourselves and define who we are. Nature has categorized animals into species and we categorize, or label ourselves as this or that within the human species. Maybe homosexual identity is partly just a byproduct of consciousness.

It seems this categorization has taken on a more serious tone under the modern religions. Among ancient polytheistic groups and even more recent tribal peoples, homosexual activity was used as a means to regulate population growth. But this stood in the way of the strategic population growth of the modern religions. Once the major religions put explosive growth in their strategic plans, homosexual activity became taboo and was outlawed in the religious scriptures. Arguably the most important factor in the Christian rise to power was their sheer numbers. As we mentioned in the section on beliefs, behind every good belief is a GPR boosting strategy. The Christian strategy was to grow, and the beliefs sewn into their scripture led to massive numbers of Christians. "Go forth and multiply", they were told. And homosexuality, along with masturbation, abortion, and yes, bestiality, was a bane to this.

With homophobia now in the scriptures, the label of homosexuality became more laden than other labels we might give ourselves. The ban itself may have been the push that started homosexual identity in the first place. Any individual that had homosexual tendencies would have been forced to take sides – *with us* or *against us*. Sexual preference being so hard to deny, many would have simply had to choose for their homosexuality and bear the brunt of discrimination. No more living on the homosexual spectrum, sexual preference would have become black and white.

My postulations on the birth of homosexual identity as a thing are simply that: postulations. Maybe sexual identity became important once homosexual activity was outlawed. And maybe homophobia came along with that like a pair of headsets with a smartphone. Maybe it became dangerous to be thought of as gay and we worried about ourselves being accused, but also worried about the sexual fate of our children. And maybe homosexuality got lumped with the other vices, pedophilia and bestiality, simply by human nature. You see one list titled "Good" with one item: sex with your wife, and you see another list titled "Bad" with a short list of the other varieties and you literally lump them together and don't make the effort to distinguish.

Studies actually find that homophobia can be linked to fear of our children turning homosexual. Author Jesse Bering quotes a 1983 study by

Gallup and Suarez which states "*So-called homophobic reactions should be proportional to the extent to which the homosexual [is] in a position that might provide extended contact with children and/or would allow the person to influence a child's emerging sexuality.*"[33] What they are saying is that we are most repulsed by homosexuality when it might involve our children. Being the genetic investments that they are, we want our children to grow up and have children of their own. To that end, parents want to keep homosexuality out of the picture as much as possible.

This may be that lumping of sexual vices since pedophilia and homosexuality aren't the same. Homosexuality is being attracted to members of the same sex. Pedophilia is being attracted to children. The link is illogical. But then, much of what we think is illogical, but we think it out of convenience and because being wrong hasn't hurt the 90 or so percent of us that is straight.

It is also hard to say if this way of thinking was present through our evolutionary history. If so, then it will not go away any time soon. If not, then the challenge still remains for homosexuals to distance themselves from the sexually deviant and be treated as equals by a heterosexual community that sees everything outside of traditional marriage as a threat to society.

The next reason for homophobia brings us back to parasites. Anal sex, and promiscuous sex for that matter, is more vulnerable to sexually transmitted diseases than is vaginal, monogamous sex. When HIV broke out in the 1980's it was nicknamed the gay disease when it seemed to only affect homosexual men. We soon learned this not to be true – that any sexual encounter could transmit HIV – but it did show the correlation between promiscuous sex and disease. Is this just a phenomenon of modern times? No, say the researchers. This goes way back. Researchers conclude that sexually transmitted diseases helped push our species into

---

[33] Article Jesse Bering: https://blogs.scientificamerican.com/bering-in-mind/natural-homophobes-evolutionary-psychology-and-antigay-attitudes/

more monogamous sex[34]. These researchers were not looking into homophobia, only into the effect of STDs on our sexual patterns, but they still showed that our genes did better when men and women pair-bonded and scrapped the orgiastic lifestyles seen in our cousins the bonobos. Promiscuous sex is not anal sex, but they share the same correlation with increased infection. The same distaste we may have for homosexuality may also be shared for heterosexual men who visit prostitutes for the same reason. And that may explain why heterosexual males don't want to be thought of as gay: just one homosexual encounter could result in infection. The genes' strategy would then be to play it safe and not only proclaim heterosexuality, but also proclaim disgust for homosexuality.

Homosexuals have been oppressed for ages and the suffering they have incurred is horrific. So, this is a touchy subject and I'd just like to confirm once again that I don't condone or support this behavior. I'm just describing the trend and trying to explain why it exists. And it does exist. Some men are more progressive and don't care about being thought of as gay. Some cultures don't mind the idea that a sizeable portion of their population is homosexual. But when we see this dichotomy of male sexuality, we want to explain it. It doesn't exist as strongly in women. Unlike men, women often do have periods where they are lesbians only to revert back to a heterosexual lifestyle. Why can *they* do it? Again, maybe because lesbian sex has lower disease rates than does sodomy.

In the late 80's my father had a colleague who contracted AIDS. My father invited him over to our house once for a drink and beforehand he explained to us that the man had AIDS, but that it was not something to be worried about. He came over and I shook his hand, knowing that there was no danger. Our intellect can tell us there is nothing to worry about. Our manners say shake hands with the man. But our genes tell us not to take the risk. And there is a slight reflex not to get in contact with sick people – irrespective of what disease they have or how it can be

---

[34] http://www.nature.com/ncomms/2016/160412/ncomms11219/full/ncomms11219.html

contracted. Again, not saying that's a good thing, just saying that's how we're hard-wired. If there is a natural, historical correlation between homosexuality and disease and if the genes of heterosexuals that were homophobic were selected more often than those that weren't . . . well then, we become a species of homophobes. Homosexuals and homophobes, that is. Because, as we just read, some people do just fine as homosexuals too. Just not the majority of people.

It's all a bit confusing and politically charged. If homophobia were natural we would expect to see it everywhere and all the time. But this is not the case. The Kathoey (transgenders) of Thailand seem to have a decent spot in Thai culture. I mean, they don't have desk jobs but they don't seem overtly oppressed either. And what about the Greeks and more modern tribal societies whose heterosexuals did engage in open transient homosexual behavior? It seems like aversion to such things only came about in the past 10,000 years or so. Historically you'll find that homosexuality has been acceptable in some cultures at certain moments in history. This could show that homophobia is too recent a phenomenon to be hard-wired in our brains and that there may be some cultural slack that would give it a pass.

Either way, what is hard-wired in our minds is that we still live in a tribe and being gay in the Sweet Spot meant not being optimally geared for high GPR. This is no longer the case, but our minds are stuck in an older world where reproductive sex was the paramount strategy for passing on our genes. Our biological programming tells us *that we will be most fulfilled when we have children.* That's not an argument against homosexuality. But if we conclude that the goal nature has given us is genetic proliferation, then the genes for homosexuality are not ideally equipped for maximum GPR.

These are not pleasant things to say but we can't explain life and make everyone happy. Homosexuality is a phenomenon to be explained like anything else. And when explaining anything we want to be as objective as possible. If we conclude that homosexuality is not optimum for gene proliferation that doesn't mean we harbor any ill feelings to homosexuals or discriminate in any way – no more than we would discriminate against anyone who has qualities that would make them less optimized for gene proliferation. I mean, what a silly thought. I'm sure you think it's silly too, but sometimes you have to explain objectivity to silly people.

And for the record, there is aversion to all kinds of normal sexual behavior. You name it: slutty women, chaste women, promiscuous men, and chaste men. They all get a bad rap somewhere. Sexuality is as much a reflection of your essence as anything else, and you will appreciate others who share your sexual preferences and feel distant from those with differing sexual preferences. So, if about 90% of the population is heterosexual, you can expect a good bit of complaining about homosexuals. But my point is there will be complaining about all kinds of sexual and non-sexual behavior.

At the end of the day, we're all human beings and we all want a good life. Homosexuals are finding more and more room for that in modern society. Despite more shocking trips back to the 1800's in the news, they are getting official rights the rest of the free world has and are able to have a family all while maintaining their homosexuality. Adoption programs and surrogate motherhood are giving them the fulfillment the rest of us are feeling when we have children. They're able to partake in that timeless institution that makes our lives so exhausting and so fulfilling at the same time: the family. If GPR were a product at the store, the *family* would be the vehicle we all hop into to go get it. Even if adoption means you aren't passing on your genes, you're still you're instilling your beliefs and values in your children and they will do the same with theirs. As cold-hearted and manipulative as it may sound, it is most fulfilling.

Gay or straight, you want to be in a family.

Family

We all have family. We all love them on some strange level despite the fact that we can't stand them half the time. There is a unique bond there and if you've read this far into the book, you should understand it. Our family members have the highest similarity of our genes and therefore our genes have a vested interest in seeing our family members do well. Let there be love! Love, like in intimate relationships, is the contract. This mechanism was built for a different time when all we had was family.

So, yes, we love each other, but like with any relationship, we expect things. **A parent's love is not unconditional**, despite those quotes people post on Facebook. So, mothers don't need to feel guilty if they're annoyed with their child on some level. Mothers seem to come close to unconditional love early in a child's life, but the demands start to creep in once the child is older and more in control of his life. Fathers may place more demands on their children as the father complements the mother's love with a healthy dose of expectation. To do well in life, children need both, and each parent is specifically designed to add their half of the puzzle. What parents want from their children is high GPR – they expect their children to grow up to be good looking, high status adults; adults who are not only capable of having similarly high value children, but willing to do so. All indications that a child is not on this path will be met with hostility by the parent. "*Cut your hair*" or "*You're not wearing that skirt out of this house.*" No, you're not a tyrant who doesn't let your kids watch TV or play video games, or stay out late once they are old enough, but you *let* them do these things only because forbidding them would not work. I would gladly pull the plug on video games in our house, but it would simply be a recipe for mutiny. And hell, you played them too, and you turned out OK, right?

But the fact remains that in the eyes of evolution, children are a cold, hard investment by the parents. Parents need two kids just to break even in this whole investment. The lowliest, most un-ambitious people still want their kids to "do something" despite the fact that they themselves do very little. It's instinctive. I used to hate all the expectations from my parents and God knows we went into it in detail in therapy, but now I'm a parent and I expect things of my kids too. To want it any other way is to point your canoe in the wrong direction on the Great River.

And truth be told, kids also want high GPR, they just don't know it yet. In fact, they don't know anything. They have no idea how the modern world works and all their groaning can be traced back to them being annoyed that they have to work so much in life. And they think "putting on shoes" is work! School expects them to work and you expect them to behave and keep a reasonably tidy room. They don't like it. But they know they want stuff: food, cool clothes, toys, TVs and video games. They know it's better to have money than to be poor. When they hit puberty, they know it's better to be popular than a dork.

But the struggle is everywhere, let us not forget, and natural selection is that search for balance that the universe demands. **As parents manipulate their children for their own genetic gain, children also play their parents like a fiddle**. It starts with being cute. My little baby boy is so damn cute! Kittens have pictures of *him* on their iPhones! I'm late for work trying to get out the door and there he comes crawling up to me and he wants me to pick him up. "Yes, master, you cute little bugger, I will pick you up and cuddle you and protect you from anything and everything in this world to the utter detriment of myself." I am putty in his hands. My wife and I spend countless hours caring for this boy. Is it all necessary? No. Does he *need* me to hold him while I make his bottle? How about "Thanks for all this care, Dad. Love that the heater is on."? There will be none of that, but still we care for him. Is it all done for our own GPR? No. He's got his own genes and his own plans for them. He is simply using us. Yes, natural selection has even built babies to be ruthless GPR investors. Despite the love I feel for him and the love I feel from him, we compete, as much as cooperate.

And so we see that parents are just another resource to kids; one that they also try to exploit as much as men and women exploit each other and as much as shareholders exploit the earnings of their company. Of course, they don't want to clean up the dinner table after we eat. They expect food to be bought, cooked, served, and cleaned so that they can focus on doing the things they want to do. They want clothes bought, ironed, and picked up off the floor. Doesn't mean they get it all from us, but they want it and they try to get it. Children are constantly trying to get more out of parents like their parents are trying to get more out of everything else. It's totally natural.

Some of you are reading this thinking, "Not me—I have sweet kids and we have a great time together." Good, that sounds like a more cooperative relationship and things are going well for both parties involved. But behind it still lies the goal of higher GPR. If both parties are holding up their end of the bargain, then great – you'll have a hard time seeing the struggle and manipulation in all of this. But imagine that you stop doing the things you do or they stop being so sweet and obedient, and maybe you can see how things will fall apart, and maybe you'll understand what's creating the tension: the broken contract.

The kids have a similar relationship with each other. Sibling rivalry eventually came out as a concept in 1930, led by psychiatrist David Levy. We all knew it was there. You love your siblings but you fight with them too. Some of you dream of them not being there so you can have your parents all to yourself. **Your parents are a resource that you need for high GPR. Each new sibling dilutes that resource**. It's no wonder the older children resent the children under them. And no wonder young kids hate new siblings. At a young age they need their parents' resource more strongly than when they are older. We have two older boys and the level of jealousy they feel for their little brother is pretty minimal. Yes, we've slacked off a bit with the older boys since the little one was born, but they don't seem to mind too much. They're old enough to take care of themselves for the most part. But it wasn't the same when our second child was born just two and a half years after the first.

Not only is there rivalry between siblings, there is rivalry with unborn siblings. My children are master cock-blockers. As any young man knows, cock-blocking is the art of keeping another man from being with the lady you like. Now I have been cock-blocked in my day, but never to the extent or sheer level of cruelty that my own sons have dealt out. As they try to keep new little ones from being conceived, my wife and I have been relegated to the sneakiest forms of sex. It may sound exciting, but it is not. Sex is best when the moment takes you, but these boys and all the work they require ensures that there are few moments in which to be taken. My wife and I are the celibate victims of their strategic warfare.

As we've talked about GPS, we know that personality is really just a strategy for getting higher GPR. As Steven Pinker describes in *How the Mind Works*, it makes evolutionary sense for older children to be jerks while the younger children, not as strong and able to deal with physical conflict, take on a more accommodating personality. Sometimes, you'll get the opposite: a softer first-born makes it possible for the second child to be more assertive. They are both genetic proliferation strategies and both born out of the initial environment a child is born into. As parents, we hate to see our kids fight and hope for the moment when their competition turns to cooperation. We know that's where the higher GPR lies for everybody these days.

At its core, family is just another opportunity for improved GPR. When a gene is in many people, it has a better chance of getting passed along. It thereby also has an interest in the other carriers. With all the traveling and interracial breeding in the modern world, our genes are both spreading themselves out more, and having a harder time distinguishing carriers from non-carriers. But we are always grouping. We start with family and then edge outwards to a larger circle if the technology allows us to do so. What do I mean by technology? Well, many thousands of years ago you would have lived in a tribe and all of the people in the tribe were family on some level. The only technology you needed was simple recognition. Recognize someone and they are family. You help them, you love them – even if you are competing with them on many levels. See someone you don't recognize, then you kill them if they are male, or you abduct them if they are female. That was as high tech as it needed to be.

But when groups started merging to form bigger, stronger groups, a more refined technology was needed. You can't just see who's on your team and you can't remember which person your mother told you was your second cousin. You need to keep lists, organize things, allocate resources, separate and specialize people for certain tasks. And thus begins a new technology: government.

Government

> ***"I am not four people."***
> — Paul McCartney, when asked why his solo career never produced songs of the same quality as when he was in the Beatles.

You live in a country. Why? Your country has a government. Why? You vote for people to run that government based on what they say they will do. Why?

We rarely ask ourselves these basic questions. You, dear reader, are a group of cells. All your cells have the same genes so nature built you as a system in which your cells work together. Since people do not share the exact set of genes, we had to create a system to get them to work together. This system is a government. The struggle to control that system is called politics. Republicans vs Democrats, Left vs Right, Communists vs Capitalists. People create a government and then they argue about how it needs to work. The bigger the group, the more complicated the government and the politics get. *C'est la vie.*

Most people understand that we need governments but they hate politics, like it's some messed-up human invention that's long overdue for total overhaul. Daily conversations about government and politics imply that they both stink, but this is silly. Governments help us work together. They are a technology created through evolution. If governments are not there, we're not working together. Politics are how we collectively decide how to use that technology. If we think politics are corrupt and manipulative, then we don't understand nature. Politics are simply a reflection of nature, and when we understand the struggle of the genes behind politics, we understand that it's naturally a real mess.

Today we all live in countries because at some point it became better for your GPR. Being part of a country means you are able to leverage the strength of a large group. You have a government because we all need the same things and it's more efficient to set up an entity to take care of all those things than if we each spent time doing them ourselves. Your GPR is higher because you share things like roads and hospitals and do not have to set them up and maintain them yourself. Underneath it all is the genes of the people in that country. As the genes hope to fare better in the world by creating human carriers, humans hope to fare better in the world by creating governments. The genes of the people who created governments eventually outnumbered the genes of people who did not.

This all started a long time ago. Tribes merged into bigger groups and they merged into even bigger groups and fast forward a few thousand years and you have countries with formal governments and borders. Groups of people who did not set up governments were not able to work together as well, so they got conquered by those who did. It's just another technology. Then you had good governments and bad governments. **Like any other human system, governments are subject to the judgment of natural selection.** Duh! Their goal is to improve GPR and they live or die depending on whether they do or don't. History is full of civilizations that either conquered or were conquered. Often it was the government, or lack thereof, that made the difference.

Groups and their governments change with time just as animal species change with time. All along the way nature is grading them and making them change for the better. If they do not change for the better, they die. The Roman Empire did great for a long, long time. Its members' genes were living it up. But then it broke down when conditions changed, just like it can with animals. And the Roman Empire went extinct. It's the same phenomenon as animal extinction. No group exists for long which does not support the GPR of its members. If being in a group gets you killed, that's a deal breaker. If it is set up for a certain environment and that environment changes, and the group doesn't change with it, then it dies. On the other hand, if it provides you with good roads to drive on, a growth-oriented economy, freedom to marry who you want, possibilities to grow outside your country (export stimulation), then your GPR will go up as a result and the group will be on firm standing.

When we think of governments, we think of countries but there are many other groups that have governing systems. What other kinds of groups are there? Cities, states, religions, organizations, companies, etc. You are in a few of these. Some are big and small but there is competition and cooperation in each one of them. The Four Recipes are all at work. The same strategies that genes use with men and women and on all levels of sexual competition are happening on all levels of government. Yes, groups look out for each other. Big companies look out for other companies (maybe suppliers) and within those companies, departments can help each other out, and within the department there are groups of people who are on your side and some who aren't. Yes, groups are jerks to each other. We see this so clearly in current American politics where both Republicans and Democrats (OK, mostly Republicans) spend lots of energy trying to derail the competition. By now, we are not surprised. This is part of nature. If Republicans are good at being dicks (and they are awesome at it), nature is OK with that.

And yes, groups have sex. I'm talking about companies merging with other companies, and countries joining other countries. Currently, part of Belgium wants to become part of France. Same with parts of Canada. They both want to have sex with France (I know, who doesn't), for doing so will result in a recombined country with better potential than before. Or so they hope. . . . Groups have always looked at each other and thought: do we fight them or do we . . . um . . . join them.

It's all about getting higher GPR. That is what the genes of the members of the group expect. Are they getting it? Of course, not all individuals in a group are equal, so not all individuals are seeing higher GPR. Like, America has a pretty high GPR as a country, and within America there are people with high and low GPRs. Get it? This is the nature of groups – not all members are equal. It has biological roots. I mean, can you imagine a group of animals where every individual is equal? It doesn't exist. Ants and bees have a queen. Groups of mammals have a dominant male. **Human nature is not set up for total equality in groups**. There is investment and there are rewards, and government is always about who does what and who gets what. But everybody brings their own level of investment to the table and therefore everybody expects to get an appropriate reward. We all put stuff in, in order to get more out **as a group.** But now the question becomes, *How do we distribute the rewards of group living to the members of the group?*

Imagine an individual who works and makes $9000. If a group of 100 of these people work together, they can make $1 million collectively – which is $10,000 each. So, it makes good sense to work in a group. But of course, by the nature of human beings and the whole synergy of different roles and cooperation, not everyone puts in equal effort. So, some people want the rewards divided up differently. Dividing the rewards up evenly is simply not found in the animal or plant world. It is only found within individuals – *groups of cells which share exactly the same genes*! **Groups with slightly different genes never divide the rewards of collective effort evenly**. Communism tried it and it failed. "From each according to his ability, to each according to his need" won't last long. It is pushing upstream while the Great River is flowing mightily downstream. **We may like the idea of every one being equal, but nature does not**. When everything is equal there is nothing to choose above something else. Equality implies an environment that never changes and that just does not exist in our expanding universe. This is good news for the achievers and a burst bubble for the underachievers. I'm an underachiever, so I don't like it either, but, again, we're trying to see things the way they are, not how we want them to be.

So, stuff has to be divvied up in a way that keeps the motor running well. And you may have made $1 million bucks together, but you also had to invest in things like roads and laws and organizations to regulate and stimulate. How much do you spend on each? How much did I invest up front and how much did you? Should we spend $100K on roads and $50K on the legal system? Or the other way around? You may think one thing and someone else thinks another. Who gets to decide?

And remember that slight differences mean a lot. If I bring $9000 to the table and get a 11% return on investment for being in a group, but someone else brings $9000 to the table and gets back 13%, I'll be pretty ticked off. **That 2% is a big difference!** So, we are designed to be very careful about getting our fair share in life, because if we do not, our genes suffer. Look around at the politics in your country. All the commotion is caused by people fighting over that 2%. As annoying as it all is, they are right to do so! It's a big deal. And our whole being is set up to make sure that we come out on top. Combine this with the unbelievable complexity of modern life and you get what *The Daily Show* refers to as a *clusterfuck*. The political struggle is a chess match of laws and political scheming all designed to get that 2% — a chess match no different from the one playing out in the pond near your house where billions of organisms are trying to pass on more of their genes than all the others.

Spread the wealth

As all groups are designed to improve GPR, we see that political affiliations are first and foremost about the struggle to control the resources in the group and distribute the benefits of the improvement. *Do you get 11% or 13%?* **Government is *supposed* to redistribute the wealth**. Sure, it is also a technology to facilitate better working in a system. For people to work together to get better results there needs to be a manipulation of the environment, in the form of laws and technologies. Like you need laws that say "don't kill each other" and you need technologies like roads and schools. Take these things away and you have an environment which doesn't allow you to work together well, meaning you're back on your own making $9,000 instead of $10,000.

But control of those technologies and the distribution of their fruits is what it is all about. It's in a gene's interest to get an unfair share of co-operative efforts so genes that pushed a person to do this fared well in evolution. But if you feel you are getting cheated in your current group, you can leave. This is most clear in countries and companies. Some people want Texas to secede from the U.S. They could do, but would they be better off on their own or in their current group? Would the collective GPR be higher or lower? If all Texans feel it will be higher, they'll go. And we'll try to stop them if we think *our* GPR will be lower if they leave.

Perhaps you work for a company. Would you be better off working for yourself? Many people think so. There are upsides and downsides. You won't have the hassle of office politics, but you won't get the dental plan either. If you work for a company and they make a profit, as they should, maybe you will expect a percentage of that profit as a bonus. That's the divvying up of rewards. The executives will expect a bigger bonus and the lower workers will complain about it, but theoretically they did not bring in as much GPP as the executives did.

So, in countries, people on the bottom complain about the greedy bastards in power, and the people on the top complain about the free-loaders at the bottom. In companies, the guys in the cubicles complain about the executives big salaries, and the executives complain about the lack of "team spirit" from the guys on the floor. But in each case, they all want the same thing — an unfair advantage in their GPR.

Improving the technology of government

As a technology, governments are always under development and the policies they create are always changing. Slavery is now illegal but it used to be in the law books. Healthcare in the U.S. used to be optional, now it is mandatory. China used to be strictly communist, but it later opened up to allow for more free-market enterprise within its rule. These changes are a result of both the internal power struggles of a government's members and the strategies that succeed and fail. The trend has been for the people to decide what the policies are. **The current high-tech political**

system is a democratic government. Since governments are here to serve the genes, then it should be the gene carriers who decide what stays and what goes. This sounds pretty obvious, but forms of government like dictatorships and communism do not follow this ideal. The only way they are sustainable is if they actually *do* improve the GPR of the people. China is improving the GPR of its people despite being Communist. Communism in Russia didn't and it crumbled. It is obvious in most dictatorships that GPR would be higher if they would trade up for democracy, but the problem is that GPR would drop sharply if they forcefully challenged the dictator, eventually ousted him, and had to rebuild for many decades to get back to a state of stability and growth. North Korea is currently holding that bulldog by the ears. Iraq was gently nudged into that "liberation process" by Uncle Sam.

Policy = Personality

Those three wonderful acronyms still apply. There is a sort of group GPR and group GPP for the members of a group. A group has a certain value and when they do well their value rises. This was the drive we had many thousands of years ago and it's still in us today. We can also see that each group, be it a company or a country, has its own strategy for getting higher GPP. Just like with an individual, that strategy is GPS. Just like there are archetypes with people, there are archetypes with groups. Some companies are nerdy, some are jerks. Some countries are more cooperative, some are more combative. Just like with people, those strategies are genetic proliferation strategies. Holland doesn't really get involved with aggressive strategies. They have a tiny army, so they really couldn't anyway, even if they wanted to. But this changed from hundreds of years ago when they were proper bastards. Their GPS shifted when their imperialist ways stopped working and their GPR did better as a result. The same is true of the U.S. They used to be this growing country focused on getting all their land settled. Now they are a dominant world power and their GPS changed with that reality. **What we call foreign policy and domestic policy are just the GPSs of a country**. What are their strategies for doing well in the world? We have world powers, mean states, peaceful states, industrial behemoths, fat cats, corrupt, strife-ridden states who can't get their acts together, shrewd upstarts, etc. Democracy lets the people decide what that strategy is, but like most things human, it's not entirely conscious.

What democracy does is it allows the subconscious to vote. We *should* all vote in our own self-interest, because that is what the government is here to support. It just so happens that we are already built to vote in our own self-interest since, as you should understand by now, the subconscious will only let us work for our genes. If you think that's unfair to poor people, then let me show you that the beautiful part of it all is that if things are truly unfair to poor people we either a) get more of them, at which point they win the next election, or b) they get so oppressed that they create a violent society that rich people don't want to live in, thereby stimulating them to give more to the poor. What those in power should understand is that the more the poor have, the more they have to give to the rich. And that does seem to be the trend – financial inequality is on the run. That may not be true for some individual countries, but that is the overall global trend.

Nature supports inequality, but only to a certain level. As we've seen, all animals compete with each other and this competition can only be there if there is slight inequality. Gross inequality got shot down long, long ago in biological evolution. I can't say for sure, but it looks as though there is a rate of expansion in the universe and that rate is linked to the level of inequality in the living world. A high level is not sustainable in the living world – not for animals and not for us. We saw peaks in Medieval Europe and with the Incas where some people had seriously high GPR and the rest did not. But that trend has been on the wane. We now have some of the most egalitarian societies since the Sweet Spot.

Despite the idea that the U.S. is a pioneer, America is usually late to the game when it comes to these sorts of trends. The powerful in the U.S. are doing their best to keep the poor poor but it's a losing game.

The working class people don't have to be rich, nor will they be, but they want enough not to have to sweat at night. The rich would only have to give a tiny bit of their wealth away in order for everyone to be happy, or should I say "fulfilled". And that was the whole point.

But that's just my personal take on things. You have a personal take too. In whatever group you are in, you have a personal interest in coming out with slightly more than the rest. That is the nature of competition, and it's why politics are so messy. No one is simply acting in the interest of the group. You might think that you are by voting for leftist policies, but if we really look at it, we see that this is one of those faulty interpretations your subconscious lets you believe. It's OK that everyone is acting in their own interest, since that is what nature expects and this whole incredible world we have was built by people acting in their own interest.

When we see that the genes are the drivers of our political aspirations, we can understand that it's not selfish interests running governments, but gene-ish ones. When we act in a group we are looking out for each other because that tends to work out well for our genes. As Sting said "We hope the Russians love their children too." So we seek out ways to create a better world for all of our children. And as we've seen, we want to create a better world for all those people with our genes and all those people who help out our genes – and that's a really big group of people. It's all subconscious, it's all driven by the same motivations we've had for millions of years. It's all natural and hard-wired, so we can't change it. At best we can understand those motivations, and point our boats downstream.

Economics & Work

Everybody wants to be rich. That's natural, right? As we've seen, money is a big boost to GPR. Financial prosperity is itself a measure of success, just like GPR. Making money on a group scale is economics. Simple enough, right? My kids ask me "Dad, how much money do you make?"

and "How much money do we have?" You can ask a government the same question. In fact, you can ask the world the same question. Since I have a business, each year is a little different, and it is a little bit different each year for your country's economy and for the world economy, too.

We said we can't measure GPR since we can't count up all the genes in the world at a single moment and then count them again a year later. But we can count money. Or at least, we think we can. The more money a country has, the better it is doing, but we know that money is just a substitute for GPR. **And that's what really lies under economics: it's how well a group of genes is doing**.

What is money? What is power? What is economics? Are they all the same thing? Again, these things are so basic that we never really take a look at them. In fact, we can't take a proper look at them until we understand our purpose in life. Thankfully, that is what we've been doing here. So, now we can finally have a good look at that thing some call money, some call power, some call economics, and nerds like me call GPR.

Nature wants energy dissipated. It's a friggin' law — and there aren't too many physical laws in the world. Self-replicating organisms were born and they dissipated energy, so they were allowed to stick around. They dissipate energy well when they create new copies of themselves and pass on the genes inside them. The higher the genetic proliferation rate of these organisms, the more energy they dissipate. **High GPR = good**. Billions of years later, they realized that creating bigger organisms would improve their ability to dissipate energy, so they started transforming themselves into all kinds of animals, including humans. We kick ass at dissipating energy, and in groups we do it even better. The bigger the group, the more energy dissipated in the form of high GPR. But this was clear to you already — we set up groups to do exactly that: get a higher GPR. Despite our conscious ideas about the world and beautiful quotes about spirituality, we have the same drives we had 50,000 years ago, and those are the same drives that all living things have.

Economics is just a representation of how well we are doing in getting this high GPR. Those who measure a country's success in terms of money or power are missing the underlying goal. We don't care about those things, we care about GPR. But hey, it's OK — money and power are not too bad. They are well linked with high GPR. If you look up the definition of economics in Wikipedia, you get this:

> Economics is the social science that studies the behavior of individuals, households, and organizations (called economic actors, players, or agents), when they manage or use scarce resources, which have alternative uses, to achieve desired ends.

They don't say what that *desired end* is, but we know it now. An *economy* is measured in various ways, but it always ends up being expressed in terms of money. But we also know that this is not enough. For instance, GDP is a typical way to measure the value of an economy. America's GDP is $16 trillion. China's GDP is $8 trillion. But what about all the other things a country does that make it more powerful, or give its citizens a better chance to do well in the world? Believe it or not, education isn't counted at all when measuring an economy — but we know how important it is to groups and individuals. What about countries like Israel that have small economies but that do quite well in the world and hold power on other levels? In the end, all we care about is the ability for the genes in that group to proliferate.

If we do something crazy here and stop measuring economies on money alone, and start measuring them in terms of GPP — their ability to proliferate the genes of their members — we get a new perspective on groups. We see that behind them is not money or power, but the genes themselves.

Take a step back. Imagine a group of people. It's 20,000 BC. They live on the African savannah – all 60 of them. They hunt and they gather food because there is no Burger King. They have a pretty average GPP collectively. There's nothing special about this group. One day a guy is eating an avocado and swallows the seed by accident. “Oops,” he thinks. His buddy says, “Dude, did you just swallow that avocado seed?” “Yeah,” he says, surprised. His friend yells for the other tribe members, and together they all laugh at how this guy just swallowed the avocado seed. One person jokes “an avocado tree is gonna grow in your stomach.” He is scared. But two weeks later when he finally manages to poop, the avocado seed comes out. Relieved, he spreads some dirt over the spot and limps back to the group. He goes back to the spot some weeks later and he notices a small branch growing from the pile. It’s an avocado tree. The following year, his tree bears avocados that the tribe can eat. They taste a little like shit, but they are too hungry to be picky. He knows that the tree came as a result of him shitting it out, so he advises his whole tribe to do the same: swallow the seed and then poop it out. They do. Trees start to grow. Of course, they don’t yet realize that they don’t have to first swallow the seed and poop it out, but the point is, they have implemented a technology: agriculture. They are farming.

They now have lots of fruit in the area and can always come back to it. Maybe they don’t have to travel so much, so they have lots of energy to spare. Less of their members die of starvation. They can spend less time finding food and more time creating weapons and tools. Their GPP has improved, as we’ve seen. But we can also see that they have experienced a growth in their economy. A few years ago they had no avocados and now they have seven fruit-bearing trees. In a few years they may have twice that many. Their economy is booming, but all they really care about is that their GPR is improving. They are having more kids and those kids are growing up healthier. Even if they had no extra kids, and even if the trees were all too young to bear fruit, their GPP — their potential — would still be up. And that is what economics really measures.

Fast forward 20,300 years. It’s 300 AD. Rome is “the man.” There is a very real economy to speak of. Some areas are more advanced than others. There is specialization of work. Some people are black smiths, some are tax collectors, farmers, civil engineers, construction workers, galley hands, gladiators, soldiers, slaves, chair makers, tailors, cooks, etc.

And they all have a predictable output. A farmer might raise 300 pigs a year. The chair maker might produce 50 chairs a year. Next year, the pig farmer will strive to have 320 pigs to bring to market, and the chair maker, 60 chairs. They don't strive for *fewer pigs* or chairs than the year before — only more. They might *not* have more pigs and chairs next year — but they want to. This is the human drive, the drive of all life on the planet. And the trend has always been "what man wants, man gets." What we see here is economic growth — just like the avocado trees. And it still results in higher GPP. How valuable is the Roman Empire if all the pig farmers collectively produce 50,000 pigs? And how valuable is it if they produce 300,000 pigs? How much better off are the Romans in the second scenario? With 300,000 pigs, they can feed more workers and produce more things. They can make sure their children grow up with full bellies and are able to spend time learning instead of helping to farm. Their children will themselves be in a better position to have children who are of similar prosperity. The better the Roman Empire does economically, the better the genes of the Romans will fare in the world. It doesn't necessarily mean that their genes will proliferate at a higher rate from year to year, but the *potential* for those genes to do so will surely go up.

And here we are now. It's 2014 where I'm sitting. We're much further down the line and our GPP has gone up dramatically comparatively, but not so much relatively because of the competition of all our peers who have also had their GPP go up. If we were to take a group of 1000 people from this time and transport them back to 1000 years ago with all the technology we now enjoy, they would kick genetic ass. We are much stronger than ever before and that simple thought experiment should show it quite clearly.

In all cases, in all time periods, economic measures are simply measures of how well our genes are doing. If we understand that the genes are at the center of it all when it comes to the individual, then it should not be hard to make the step in seeing that *people* are the center of it all when it comes to companies and governments. **It is the collective gene-pool**

of the people who benefit or suffer when a company or country does well or poorly economically. And it's the genes in those people who also experience the same fate as the individuals. No matter what scale you bring it to, if it involves people, it is directed by genes.

If this is a new concept to you, you're not alone. If genes are in charge and we are playing "gene detective," we need to think, "What do the genes want from the economy?" Again, we can have a look at early economies and think about what they delivered. Obviously for early man, a good economy meant concrete basic things like more food. Farming, specifically, delivered a surplus of food, which meant there were moments when people could do things other than hunt and gather. They could make things, like chairs and fish hooks. They could even make tools that made more of these things. Important things they made would have been weapons. With the technology of farming and tool making, you would think that life would have gotten a lot easier, but competition between groups was still a reality and this competition kept life hard and kept war on the horizon. A good economy provided a better chance of winning wars — just as it does today. The U.S. has a huge army because, with the largest economy in the world, they can afford it.

Winning a war provides a few nice advantages to any set of genes. First of all, it limits the chance that your own genes will be killed. Sure you will have to survive the war, but after that, you and your children should be safer in the future. Secondly, extra women are brought into the group. If you defeat a neighboring group of people, the women were usually usurped, to put it politely. The third perk is the option of enslaving the surviving men. The first point should sit well with most of you, since it sounds like self-defense. The last two are a little uncomfortable.

Slave labor and women as spoils of war were facts of life for a long, long time. More women meant a higher GPR. The end goal is to impregnate the women of the defeated group. It may sound crass, but your genes don't know what *crass* is. . . . To them, women are fertile ground to plant your seed in. To not make use of them would be evolutionary suicide. We know that chimpanzees do this. They attack neighboring groups of chimpanzees and, if at all possible, rape any women they can get their hands on. As Steven Pinker tells us in his book *The Better Angels of Our Nature*, women are often raped during the course of war and even modern day generals "turn a blind eye" to the actions of their soldiers. At least now raping women in war is officially illegal, as Pinker states. Hundreds of years ago it was part of the motivation to get soldiers into battle. Yes, we are a sick bunch. And it gets worse before it gets better.

Slavery

Slavery is natural.[35] Other animals do it too. Humans continue to do it. The way I learned about slavery in elementary school made it seem like a thing that happened to Africans around 1600 and went on for a lengthy 250 years or so before the U.S. Civil War finally put an end to it. But really, slavery has always been a part of humanity, just like rape. As soon as we *could* enslave others, we did. The Romans had slaves, the Egyptians had slaves. Africans were enslaving other Africans way before 1600 and continued to do so well past the end of the American Civil War. Statistics show that there are more people living as slaves today than in 1860. At least now, slavery is also illegal throughout the world.

Slaves originally were a spoil of war but then it got institutionalized and people actually set up the infamous slave trade where they didn't need to wait for war to enslave people. Slavery is a technology. Those who owned slaves did very well for a long time.

Today we're all horrified by the idea of slavery, but to understand it better, we need to take out the emotion for a second. What if you could own creatures that would do work for you? Like a horse? Humans have enslaved horses for a long, long time. Yes, if you own a horse, you are a horse-slave owner. It's slavery, because even though horses get fed, they have no say in the matter. They don't want you on their back. And they

---

[35] Let me go practice saying "I was quoted out of context".

don't want to plow your fields. Worse, they are bred for those specific functions.

What if you could own humans? Imagine for a second that they feel no pain and that they have no drive for their own freedom or genetic proliferation. They're cool with whatever you want them to do. *Whatever you want them to do* . . . Would you own some people? I might. Now, what genes have done is not worry about what the slave cares about and just own them anyway.

Slavery is a valid strategy for genes. Through the course of human history, it has proven an effective one. It is so effective, it still exists despite our complete repulsion to it and its breach of international law. Whatever output you are able to create, slaves will be able to increase that output and thereby increase your ability to proliferate your genes. "Nice work," says Nature, and she gives you a star sticker on your paper.

Just to make sure some self-righteous troll isn't misinterpreting what I'm saying, let me be clear that I am not condoning slavery, I am just explaining its existence. Now let me explain its demise.

I've just called slavery a technology. It is. We are apes, but no other ape enslaves members of their own species. We invented human slavery, just like we invented fax machines. Get it? But at some point fax machines became obsolete and were replaced with a better technology. And so it was with slavery. The countries that still have illegal slavery are the same countries still using fax machines. They do not have a high enough level of technology to put an end to slavery. But they will, eventually. Being such an effective strategy for genetic proliferation, getting rid of slavery is a major accomplishment. So, hats off to humanity for getting this far. The writing is on the wall to have it wiped off the earth soon enough . . . "soon" for evolution, that is.

Technology will make it happen on two counts. Firstly, technology is more valuable than resources. And that's what slaves give you: resources. Slaves don't innovate, they perform manual labor. There came a point in the economics of the world where that was not enough. A resource-based economy will lose to a technology-based economy because technology evolves more quickly. That is the entire premise of this book. Good economies are technologically advanced economies. Good people are technologically advanced people, as we'll get more into later.

Secondly, the set-up of the modern world makes slavery more expensive than its competitor, trade. It is cheaper to buy work off people than it is to enslave them. In most parts of the world, slavery is now a poor investment. That sounds like a very cold argument against slavery, but it is all our genes care about. You will not sell them on emotion or vague concepts like morality, only on GPR. War, rape, and slavery are all technologies, they all grow the economy, and they are all being replaced just like your Blackberry. They are being replaced by trade.

These days we consider trade the only real option for growing our economy. How's that for progress? Of course, the powers that be still know that war works too, and every now and again the possibility of war is brought to the table disguised as something else. But we all know that every modern war has economic gain at its core. Trade, too, is obviously economic, but it's a way of getting what you want without having to risk death in the process. Wars aren't cheap, not in terms of money or loss of GPR (i.e., your hide), so they should have a big payoff. Trade is also not cheap — you have to make/do stuff to trade. But the beauty of trade lies in the fact that it is not a zero-sum game. If I need food and you need a horse to ride, we can trade and both be better off in the deal. And so we grow economically.

With economic growth, we as individuals and as a group, can work on not only having more children, but on improving the GPP of those children. A couple of ways that a good economy helps children are:

1. Improved health — we often don't realize just how easy it was to get sick and die in the past
2. Education — life is all about technology and education means teaching and learning technology.

Trading works best when one person has or does something and another person has or does something else and they each want what the other is offering. If one person grows apples and another raises chickens, they can trade and each person feels they got a good deal. "You want apples? Really? That works out perfectly for me, because I have a ton of them." **In this sense, the more we trade, the better off we become**. We always trade things we value less for things we value more. So, to keep this train running we need two things:

1. More stuff to trade
2. More people with stuff for us to trade with

The bigger the economy and the more valuable people in your country, the more you can trade. Every economy wants people in it — even foreigners. The real reason people dislike foreigners is either: 1. They have nothing to trade, or 2. They have a lot to trade and will eventually become your boss by making better deals than you will. In business, there is always the drive to seek out more people to trade with as opposed to trying to stay in a small group.

This is all done for the sake of higher GPP. When you start trading, you get more, and the more you get the more you can innovate so as to have better stuff to trade. It's a spiral pushed upward by trading and technology. The more a country can innovate and do whatever they are doing better, faster, cheaper, the more money they should make through trade and the better their citizens' GPP will be. Technology has become so important, it is, in many ways, considered more important than the natural resources a country has. Countries are considered to be "cursed" if they possess lots of natural resources like oil or coal, because it makes them lazy in innovating — like a good looking guy or girl who has never had to work on their personality because everyone treats them well. **Natural resources can be depleted; technology can't**.

In today's world, the most valuable resource a country has is its people. The people of today possess a huge amount of technology compared to the people of just 100 years ago. Aside from whether people work harder or longer hours now or in previous generations, one thing is for sure, they produce more now. To be able to produce more, you need to be able to know more, you need to be technical. Marketing, sales, business theory, manufacturing processes, management information systems, agricultural science, financial strategies, etc. are all technologies that make the citizens of one country do better than the citizens of another country. All over the world people work. But the people who employ technology produce more out of their work than those who do not.

This should show that technology is the key to a fulfilled life — not just for you, but for your country and your family. But if you want to sign up, you have to be skilled. You have to know how to use it. You have to know how to create more of it. And that means education.

Education

I didn't like school. I'm not gonna say I hated it, but I found it a real struggle to show up every day and manage some sort of competence. "*What the hell? Why am I here?*" "*I'm never going to use this stuff.*" In hindsight, I was totally right. School wasn't for me, but education was. **In my new understanding of the world, I see that education is human life itself; it's my life**. The quality of my life rests on the quality of my education—not school, but education. If you care about life, then you must care about education. If you understand life, then you see education everywhere.

In any system, there is an input and there is an output. People develop technologies, and rule the world with them. That's output. Education is the input you need to create things, develop technologies, and rule the world. You can't output without first getting input. And life, both in the biological sense and in terms of success in society, is about output. The more we output, the more we contribute to the world, thereby raising our own GPR. The more educated you are, the more you can output.

I'm not putting down school; it is what it is: A machine to educate the most kids in the most effective way possible. There is a set way of doing things which aims to make the most progress on the most typical kids. If you are not a typical kid, you will have a hard time in school. It's just not made for you. You probably just have a different way of thinking and learning than what's being offered. Even really smart kids have a hard time in standard school and may eventually have to be moved to a different school that can better deal with their special needs.

Education is arranged on a national level. If you're lucky, your country forces you to go to grade school. If you're not lucky, you are in a country where you probably need to work to survive. School is a luxury you can't afford. But your country *wants* to send you to school and your parents want you to go to school. An educated person is a person with a higher GPP. We instinctively revere smart people, as they have become more effective and more important throughout the course of human evolution.

People talk about the "knowledge economy" as if knowledge is something you can trade like metal and fish. It is. Countries want smart people working in their borders and contributing to their culture. Countries with smart people can innovate better, thereby finding new opportunities to bolster the economy. A country has a value, reflected in its currency. If they strike oil tomorrow, their value will go up. If they could wave a magic wand and turn all their adults into PhDs, their value would also go up. Countries always, always have programs to attract well-educated people. In Holland, if you can prove you came for specialized work, they'll give you a big tax break. The more educated the population, the more potential output in terms of GDP.

Families want smart people in them — either by giving their kids a good education or hoping their kids marry smart people. Don't you want your son to come home with a smart girl? Don't you want your daughter to come home with a smart guy? Given the nature of men and women (flip back to the previous chapter on sexual strategy), smart women are not traditionally valued as highly as smart men, so guys can be excused if they bring home a sweet, pretty girl who is not all that academic. But all things being equal, we like the educated suitors. Aside from partners, every family wants their children to be more highly educated than they were. Sure, your parents will be happy if you are simply successful on some other level, but education is often the standard you're measured on before you get into the real world where different rulers are used. The more educated *you* are, the higher your parents' GPP will be.

Teaching and learning are to culture what sex and birth are to biology. Even if you hate school, you still love to learn, maybe just different things than what you're learning in school. Nature made it feel good, like it made sex feel good.

And you love to teach. I love to watch older kids teaching someone younger. They feel big and important. It feels good to teach in a very natural way.

But beware of being taught. There is always a price that comes with learning from someone. There is always an expectation. Teaching requires energy. It is an investment and your teacher expects a return on investment. When someone teaches you something technical – like how to do something – they are assuming you are "on their side" and are thereby helping to improve your GPP which will add to their GPP. Not only do they benefit when you are smarter, they also want credit for educating you. "Cool trick, Johnny," all your friends say. Then you say, "I taught him that," just in case he fails to mention it. If someone is "not on your side" you'll hesitate before you teach him something. Or if he fails to give you your due credit, you may let him know you're annoyed or just stop teaching him all together.

Teaching beliefs is a bit different from teaching technical skills. When you teach someone beliefs, you want them to ally with you. This can even apply to history since so much of it is open for interpretation. I have a conservative friend (yes, you, Karl), who tries to convince me how awful Obama is and how important conservative ideals are for America. What he wants is for me to join ranks with him and join his group, and thereby improve his GPP. "It's not just me who thinks Obama is the worst president ever! Ask John Khoury." No chance, Karl!

Education is making use of the opportunities our brains provide. Again, same end game here. Since we're trying to improve our GPR, we want to learn, but we also want to teach. For me, the beauty of recent decades is the focus on learning more about how to be a successful human being. What are the important topics involved in this?

- communication
- financial intelligence
- emotion management
- knowledge of life

This is really what the self-help industry focuses on. You may not always be happy with the quality of the material it's selling, but you have to admit the topics are important. I spent my career at school feeling like I was smart, but unmotivated to do well, and that I was learning all kinds of things that didn't matter. They may have been interesting but they didn't matter. Unfortunately, I was alone in that feeling, so I was forced to shut up and crawl back in my hole.

One of the biggest shifts in my life occurred after school when I realized that I could learn the things that did matter on my own. I could be good at anything by educating myself. Before that I understood "I can be good at math by taking a math course," but I didn't understand that I could be good at life by studying and practicing life. It never occurred to me. It wasn't a part of school, so I didn't think it was a subject. The world opens for you when you see that this is not true. You can learn how to find a mate, have a harmonious marriage, communicate better with your kids. These are not subjects in school but they *are* subjects in life, and with the wealth of information out there and with the demanding public justifying the self-help market, you can pretty much bet there is a teacher out there for you with a method that will get you where you want to go.

Wars & Culture Wars

My kids are scared of war. We live in Holland, one of the most peaceful nations in the world. But 70 years ago, the Nazis were here and the stories are still being told by the old folks. Scary stories. No, my wife's family weren't Jews, they weren't sent away to camps, but they experienced the war. People they knew were killed. My wife's grandfather was sought by the Germans and had to hide under the floor boards. Crazy things like that. It's enough to give the kids nightmares and make them worry that it will all happen again. World War II is still hard for me to wrap my head around. I really like German people, and Europe is so damn pleasant and beautiful, you can hardly imagine it was a war zone just 70 years ago.

If you know your history, and I know you do, you know continental Europe was pretty much always a war zone. Except for the last few, every generation of Europeans has been involved in war on their home soil. That's like 200 war generations versus 3 peacetime generations. The rest of the world should be so lucky. Have a look at the current conflicts going on in the world:

http://www.quanology.org/list-ongoing-military-conflicts-world .

America's last big hometown war was the Civil War, but we had the draft for WWI, WWII, The Korean War, and Vietnam. The first war-free Americans were born in the '70s. Hey, that's me. . . . I only ever had one friend whose dad was wounded in Vietnam. I guess he didn't make the cut-off.

We see war, read about war, and think "why?" Why does this have to happen? Why can't people just settle their differences on some other level? Doesn't everyone involved realize how much stuff is being broken compared to how little can be gained? Isn't there a better way? For the most part, no, there isn't. For, as you've read, if there *were* a better way, we'd have done it for the past 100,000 years. We don't make the strategies, nature does. And she likes a good fight.

If we evolved, and our behavior evolved, then it follows that war evolved. You can take war back to fights between bands of early Cro-Magnons, then between groups of people on the African Savannah 50,000 years ago, then progress to fights between cities (once they were created), and then between empires and then between countries as we have today. Wars have always existed. There has never been a time without war. It was as much a part of life for humans as hard work and sickness. It is only in recent generations that this has changed. We are super lucky. This is what I tell my kids. "Go to bed with a huge smile on your face. There is no war going on here and it looks like there never will be."

So, is war over? Not by a long shot. There is still war, but it's going out of style. I said there's no "better way," but better way to do what, exactly? I'm sure you know by now: no better way to spread your genes. Earlier in the book we summed up Four Recipes: four things genes can do to get a higher proliferation rate.

- **Sex**: They can recombine with other individuals with good genes
- **Family**: They can recognize themselves in other individuals and spend a bit of surplus energy to help that individual proliferate his genes. They also call it *kin selection*, which basically amounts to parenting and helping out family.
- **Sabotage**: They can be pricks to other competing genes.
- **Cooperation**: They can work with other genes to get passed on.

War tends to involve all of these.

For simplicity's sake, let's say that countries go to war. Of course, it doesn't have to be *countries* that war, it can be any group — be it a band of thugs, or even a "coalition of the willing." As a competitive strategy, genes improve their GPR by killing off other genes that would otherwise steal their mojo. Countries can defeat other countries and thereby gain the resources of that country. Why else fight, right? Why put your own life in danger if there is not some big payoff? We were all right to think that the U.S. had its sights on Iraq's oil when they went to fight there. **There is no such thing as a moral war**. It's as fake as unicorns. Even the American Civil War, which we romanticize as a war to end slavery, was really about resources. It was only when South Carolina stuck up its middle finger and said, "We're leaving and taking all this stuff with us" did the actual fighting, actual risking of lives and ***genes*** begin. Sounds less than honorable if you put it like that. Abraham Lincoln himself prevented his son Robert from fighting in the Civil War. I'm sorry to take away the fantasy for you, but it's just that if you know a guy and all of a sudden you hear that he did something totally out of character, you would doubt that it really happened. I know genes, and starting a moral war to free the unrelated genes of another group of people seems totally out of character. Remember the gene detective? Does starting a moral war sound like something our genes would do?

Genes are here to proliferate, not maintain good and conquer evil. These things mean nothing to a gene — like altitude means nothing to a lead pipe. Genes will only go to war if there is a payoff resulting in a higher GPR than if they stood there and did nothing, or worse yet, got *their* butts killed by an aggressive nation. Yes, wars are also fought as a way to get a leg up on your rivals — like when a fist fight is inevitable, you might as well be the one to throw the first punch. Your rivals can't attack you if they are all dead, so let's go kill 'em first. Not everyone wants to be in a war, but no one wants to lose a war, so starting a war has been a go-to strategy for not losing. There is incentive for war on both ends. Warlike behavior was selected by evolution as a good strategy for high GPR. The groups who had genes that played pacifist got killed off by groups who saw what they had and decided to take it, or who just saw them, felt threatened, and killed them in an act of preemptive self-defense.

Despite Tolstoy's book *War, What Is It Good For?* – Robert Wright points out that war usually unites two groups of people. Germany was a bunch of tribes for a long time but it was war that finally got them unified into a strong country. It takes war and death to get groups to unite. The general rule is: the more people working together, the more progress is made, the more they are separated, the less progress. War often results in that first unification that people need to enjoy a higher level of technology. If we look at North Korea today, we see a country not *involved* as well as it could be. South Korea is doing great and North Korea could be doing just as well. It's a "hole in the market," an opportunity that the genes won't let slip by. A wasted resource must be rectified and we know that it's just a matter of time before their regime collapses. . . . Will there be a war? I don't know, but something drastic will happen. Could be next year, or 100 years from now — they are both considered "the near future" as far as evolution is concerned. Will many people die? Maybe. But the one thing we know for sure is that eventually the GPP of the whole world will be higher for it.

Still, we all want war to go away. We now know what it's like to live for a few generations without it, and we're seeing what long periods of peace can bring. To stave off war, we need to create a global environment where it becomes a less favorable option. This means removing the incentives for war — incentives that only the male subconscious will recognize. It is men who start wars, of course, and by now we know what that subconscious wants: resources and women.

Those incentives are no longer worth the risk. Men still have the drive for aggression but we also still have natural selection and she is weeding out countries that aren't flexible enough to put war on the shelf. Why is that exactly? Ladies first. . . .

Most societies today are monogamous. I don't mean that we each have one sexual partner in our life, but that legally, most countries only allow a man to be married to one woman at a time. This didn't used to be the case. In the Sweet Spot, and at many moments thereafter, men at the top took multiple wives, leaving some men at the bottom with none. The birth of democracy was the death sentence of polygamy. Note how once the European countries switched to democratic governments, their lust for war started to wane. "One woman per man" was a part of the deal and now all of a sudden the soldiers are not so into fighting. They now had a woman at home whereas before she was a concubine to the king. A man with a wife and child will not want to battle unless he believes the enemy will kill his family if he doesn't.

If a man is single, i.e., has no woman, it's not like he doesn't have options in this day and age. Hell, you can import a woman from a developing country like importing a car. If you live in a developing country, you have the ability to travel like never before, and to meet new women in new places. Women are everywhere and the only thing stopping you from getting one is your foot in your mouth and that goatee. You can sign up for dating sites and take flirting courses. What you won't have to do is fight a war. Add to that the fact that women have more control over government decisions and, since war has become a democratic decision (I know, I know, but technically it is) there is a lesser chance of wars being waged just because women aren't into them.

On to resources. As scared as we can all get over shortages of drinking water and famines in some African countries, there has never been better access to food and water for all the world's peoples as there is today. If Africans are starving, it's not because there is no food, it's because corrupt politicians are blocking its delivery. There has never been less incentive to choose war as a way of getting stuff. We live in a global economy, man! You can have anything you want if you have the money to buy it. And we've got plenty of money to at least buy the necessary things. I'm currently worried about how much this summer vacation in Italy will cost (mostly because I don't want to drive too far and we want our own private swimming pool). Are you serious!? This is the plight of the middle class in Western Europe. Ain't no one going to war over a swimming pool. Fact is, even with the economic crisis of today, most countries are still doing pretty well compared to where they were in the '60s, and war was not even on the table in first world countries back then. Even if we want something and can't get it, it's not usually worth dying on a battlefield for. The stuff we need, we have.[36]

War is a zero-sum game. One side loses and the other side wins, and even the winning side incurs losses. For GPR, war doesn't make much sense today and it is losing favor with nature as a valid strategy. Compare that

---

[36] So, why did America go to war with Iraq and Afghanistan? Great question. I'm not qualified to answer that. Maybe trade wasn't working out and some Washington bigwigs thought it would be more cost effective to just go in and take what they wanted. We at least know the decision was made by people profiting from the war but not fighting in it.

with the positive outcomes both sides can have when engaging in trade. They give us bananas, we give them money to pay for education. Sure, they lose bananas and we lose money in the deal, but they have too many bananas anyway, and we have too much money, so really, no big loss for either side. Trade is the best alternative to war that nature ever thought of.

Culture Wars

War is going out of style but that doesn't mean our genes aren't fighting anymore. You better believe they are. But just as genes went from succeeding through purely biological means to cultural means, **genes also fight wars culturally now**. You may have heard the term "culture wars." It's not some buzzword for the cable news channels and best-selling pop-politics books, it's very real.

Wars are about high GPR — making our genes more populous in future generations than they are now. You can stop fighting physically, but if you want to maintain your place in the world, you have to keep fighting culturally. And whether you know it or not, you are. How does that work? Remember that cultural evolution has amounted to all the technology we have now in the form of language, knowledge, and all the technical marvels we have today, like roads, cars, phones, refrigerators, etc. All these things improve our GPP, which is simply the potential for us to get our genes into future generations. We also use culture, our thoughts and beliefs, as a way to join people, to get them on our side in our quest for genetic proliferation. If physical wars are a means to kill off other genes and find new vessels to plant our genes in, then culture wars are ways to get new people to ally with us in a cooperative effort to spread our genes, and to kill off other people and groups culturally. We saw this in the chapter on beliefs. People group together by beliefs and then work together in gene proliferation.

Culture wars are fought on the basis of all sorts of ideas. Coke vs. Pepsi. Apple vs. Android. Gun control vs. Our Second Amendment Rights, etc., etc. **People will always find ideas which will facilitate the forming of groups**. And people can be involved in multiple groups. I'm totally into Green Day at the moment. There is a whole legion of Green Day fans and somehow, someway, I am in a group with them. If I ever meet one, I'll buy them a beer and recommend them for a job. Lady Gaga seems to have successfully created a cultural army – millions of people who love what she does and have aligned themselves with her work. They will do favors for her, above and beyond buying her songs and concert tickets. These kinds of cultural alignments create power shifts in our society, just like wars create power shifts. Those millions of fans have resources which used to be scattered in all different directions, but now they are aimed at her. Pretty impressive.

All people want to tilt the cultural pool table in their favor and are actively working to tilt it. All groups want this; be it countries, companies, rock bands, you name it. All these groups are ultimately run by people, and those people are ultimately run by genes.

How do they do it? How do genes get people to ally with them? How do ideas fight with each other? And what do winning ideas have to do with genetic proliferation? The obvious example is seen in religion. Instead of Christians and Muslims today going out and actually killing their competitors (though both sides, especially Muslims, are still considering this option) they engage in a cultural war. **They don't fight; their ideas fight**. Indeed this is much safer, but the bottom line is still the same. Both sides still want the whole world to be populated by people who hold their beliefs. This is not strange, not evil, and not unnatural. It's what we all want subconsciously. Our genetically driven subconscious brains push us in that direction. This translates into people evangelizing their beliefs and bringing up the inadequacies and faults in other beliefs. "Look how great it is being a Christian" or "Muslims will prevail." They're like sports teams (another cultural entity that competes for your allegiance). The more people who latch on to these cultural groups, the better standing they have in the world and the higher GPR they will expect to see as a result of it.

Look around you as you go out into the world this week. Pay attention to the lines being sold to you by companies trying to get you to buy their products. Their ads might not be directed at you specifically, but they are branding. Branding is nothing other than fighting for space in your head. If there was nothing in your head already it would be easy, but there is lots of stuff in your head and new ideas will have to struggle for resources like a fish dropped into an existing pond. They will have to fight a little war to get in there and stay there. Apple and Android are two ideas at war. Yes, we think of them more as companies or brands or even products, but what ends up staying in your head is an abstract representation of each — a thought, an idea. Apple wants your head filled with happy Apple thoughts and Android wants the same. When either achieves this, you will be more likely to buy their products.

This is basic marketing, but Quanology wants you to take it a few steps further and see the genetic drives behind it. When Apple wins the culture war, Steve Jobs gets a higher GPR. Steve Jobs may be dead now, but his genes are not — he has kids and relatives, so his GPR is still alive. That's end to end. In between, we see this: Apple wins against Android in your head, you like Apple, you buy Apple products, Apple does well financially and gains trust which, in turn, gives Apple better access to that space in your brain for related ideas. Steve Jobs gets rich and Apple gets powerful. Steve Jobs' GPP is high; his children are well cared for and have a great chance at higher GPR not only from the wealth he accrued but also from his fame and status. That's really all there is to it. It's so mundane, but it's happening everywhere on all different scales. Most of us are not as famous as Steve Jobs but we don't have to be; we still use the strategy and get the appropriate gains. This same pattern can be seen in divorced parents who try to get their kids to take sides. It's in pee-wee soccer where the teams strive to win and get higher status in the community. Look around in your life and you'll see it everywhere.

We all want our thoughts in other people's heads. This is a way to improve our social network resource. We're all going to die someday but I don't think people at a certain age are afraid to die anymore, just afraid that no one will remember them when they do. This is what we really mean when we talk about "our legacy." We all want to be remembered after we die. It's our little way of implanting our image in people's heads as a thought that can be passed on. "My dear sweet mother, God rest her soul" wants people to remember her fondly. The real agenda here is that if the world remembers her, they will continue to support her kin after she is no longer around to do so. It's your social network resource gone eternal. Sounds far-fetched but looking at extreme examples helps shine a light on the trend that is also present in more average people.

Remember those charts I showed you where you don't even see the difference between the compounded and non-compounded interest because it's too small and the lines are too close together? But when we zoom out to 1 million years, you can suddenly see the difference? That's the point here. An extreme example can show the common trend more clearly, just as natural selection sees it.

Think about John Lennon again for a minute. He's one of the most fondly remembered people of all time. How has his family been doing since he died? I don't know about all of them, but I would imagine pretty well. And the degree of benefit would be most felt by his direct kin and then fizzle out as the relatives get more distant. Yes, his first wife Cynthia had an awful time after their divorce, but it would have been worse if she was never married to him in the first place, or if he was never the famous man that he was. It may have been better emotionally for her, but her genes don't care about her emotional interpretation of things, only how her GPR did. She did get money and she does get access to opportunities that most divorcees don't get access to. She even had her own talk show for a while on British television. Likewise for his son, Julian. He had a career in music after his father passed away. Sure, Yoko and Sean were left with millions after he died, so we expect them to be doing well for financial reasons, but they have also inherited his fame and social network, something just as powerful as money. Imagine (no pun intended) where they would all be now if no one had any recollection of John Lennon. It helps to be remembered as a valuable person. Your status, or more accurately, the memory of your status can be passed on like a good set of genes through cultural evolution.

Personally, if I could go back in time and change one event in the world, I would stop John Lennon's assassination. I wouldn't kill Hitler and I wouldn't stop Stalin from being born. I would put my own hands around Mark David Chapman's throat and stop the whole horrible deed. How unreasonable is that? That's how powerful cultural ties are. I want John and Julian to reconcile. I want Paul and him to be buddies again. More than anything, I want to hear what he has to say about current events. I don't know any Jews who were killed in the Holocaust, but I know the Beatles intimately. They're family — my family, and I want Uncle John back.[37]

"This guy told me his entire life's story . . . against my will."
- Ron White

This should show us the link between the physical struggle of biological evolution and the mental struggle of cultural evolution. In the biological world, it is the genes at war, trying to dominate a physical environment by using their hosts to fight physical wars and have sex and raise the babies. In the cultural world, created by the genes, it is the thoughts and information at war, trying to secure the minds of physical beings who would aid the creators of those thoughts and information. In the biological world there is physical violence and sexual aggression. In the mental world, we have culture wars.

Everywhere and all the time, people, companies, and governments, are trying to impregnate you with their thoughts. This is what we call *brain washing* and *indoctrination*. Those are heavy versions, but there are also more subtle versions, like beer commercials and music festivals. As I've told you, the Beatles knocked me up good when I was only

[37] Oh, don't worry—if I ever got the chance, of course I would kill Hitler. It's just my emotions talking.

12. All these people, all these groups, want you to be carriers of their thoughts just like all people want you to be carriers of their genes. Not consciously, of course – they're not aware that that's what they're trying to do. It's a natural phenomenon; an old technique that provided a slight advantage to our genes. And remember that John Khoury told you that.

War continues to happen on a physical level with actual people being shot and bombs being dropped and women being raped by soldiers, but we also see that the technology of culture has paved the way for wars of culture where ideas fight and allegiances are made. The goal of both varieties is the same. Sure, no one dies in culture wars, but genes do, slowly but surely. Slight differences mean a lot over a long period of time. . . .

Violence

Raise your hand if you've ever gotten your butt kicked. I'll expect that most guys at least have taken a couple of shots to the face in their lifetime. Raise your hand if you've ever been sexually assaulted. A lot of ladies in the house have had some kind of unpleasant encounter. War is violent, but so is everyday life for the individual.

I've personally only been punched once. Three guys followed me to a parking lot and acted kind of friendly, asked me what time it was, surrounded me, and then one of them cracked me in the eye. At least, I think he did. I didn't actually see it. I ran away. I could think of nothing else for days. That's nothing compared to horror stories I've heard from friends — stories involving knives, baseball bats, and chains. And my friends are middle class. People from poor neighborhoods have stories that blow our minds.

And then there are the rape stories. And not even rape, just women telling about the utter lack of respect and lack of restraint men show when they get close to women. Men they know, men they work with, and men they've never seen before. Talk to most girls and they will have a story or two to tell you about how a guy did this or that and wouldn't listen to any objections.

Violence has always been a part of humanity and ditto sexual assault. It's not that we're *animals*, it's that we're animals. And animals go through the same things. We should consider ourselves lucky that we even experience politeness and loving sex at all. Both physical violence and sexual assault have the same objective at heart: They both want to proliferate genes. The violence that men show toward each other is a way for a genetic vessel (you) to kill off the genes of a competing genetic vessel (him). The violence that men direct toward women in the form of sexual assault is a way to get their genes into women without their consent. People say "rape is about power," but it's not power in the sense of one person blindly trying to overpower another person, it's about a man using physical dominance to plant his seed in a woman without her agreeing. "Power trips" are just conscious interpretations of this subconscious drive. Why don't men just put women in head locks if it's about power? Cause no one ever got pregnant via a head lock.

I think this should be obvious to anyone who has not been wrongly informed otherwise. Certain people claim that "sex is about pleasure" and "rape is about violence." But if we fit rape into any evolutionary model of psychology or behavior we'll see that this is absurd. Many animals are active rapists but no one would argue the case that they are raping with the end goal of violence, let alone power. Does it make sense that orangutans rape to impregnate females, but we do it because we want to hurt someone? Even those pretty ducks you see at the park regularly gang rape females. Are they on a power trip and just trying to act cool? Of course not — they are trying to pass on their genes. Again, to better understand human drives, ignore the conscious mind for a moment, and you'll see the subconscious drives beneath. They are in control.

We know where violence comes from: It's a valid strategy for getting a high GPR in the Sweet Spot. Genes who used violence to get the resources they needed or to kill off competing genes had a higher GPR than pacifist genes. So, we have a deep innate drive to be violent. Men more so than women. Sounds pretty dismal, but there is a bright side. There is evidence that shows less and less violence in the world, despite what you see on the news. Today's circumstances do not favor violence as much as they used to. Natural selection is selecting *cultures* which favor cooperation over violence, but only because the genetic payoff for cooperation is starting to win out over violence. We see it in the most egalitarian societies, like the Scandinavian countries, who have high levels of monogamy and smaller differences between rich and poor. These two factors provide less of an incentive for violence, and hence, the flexible human brain adopts a more tempered nature. If you are from America and you want to experience something new, go to Sweden, Norway, Holland, Finland or Denmark and you will be blown away by how safe you feel. Now that's real freedom.

There are two main kinds of violence toward women: domestic violence and rape.

Domestic violence is when a man is aggressive to his spouse as a way to enforce that "love contract." You may think, "That ain't love," and maybe you're right, but as we've seen, love is just an agreement that two people have together to help each get a higher GPR. After he gets what he wants in terms of GPR, a man may become a dick, and turn this cooperative effort into coercion, forcing his female partner to do more than is fair and/or force her to remain faithful. It might also be the case that a man "claims" a woman who is not yet in any contract with him. You see women getting taken by men and they end up being the slave to that man – either for sex or simply for doing whatever he wants done. This can also come in the form of forced prostitution. In all cases, he is violently using the woman to improve his GPR – as a slave or as a partner caring for his children.

Rape is the other nasty one. To the male subconscious, in large part, women are a place to plant their seed. Men have little reason to destroy fertile ground. Woman on woman violence is also not worth the costs.

Women want to eliminate competitors too, but a sub-par performance in that competition does not result in out and out failure as it does in men. If a woman has competition, it's not a problem: The man you are after can impregnate you both. Sure, you want him all to yourself, but you won't be driven to put your life on the line for it. All you really want is that your children are taken care of, and since you are the main caregiver to your children, you need to stay alive.

But the threat of rape is always present. Not all men are the raping type but there are enough around who are not bothered by the idea of being sexually forceful. Some wait for women to be alone and then violently attack. Others flirt and play friendly in hopes to get a woman "close" to sex and then just forcibly carry out their goal. We call this "date rape." Genes want to be replicated and the ones that have done so successfully have been passed on. There is no rapist gene, but there is a rapist drive. And though we consider it disgusting, we can at least understand it. Doesn't mean we don't want it to disappear — I think we all would love it if rape and violence were wiped from the human experience. But understand it for what it is. If men are pigs, don't forget that women carry the same genes, and give birth to those men. I'm not defending it, just explaining it.

In their lives women deal with the spectrum of sexual assault. On the one end is violent rape and on the other end is "I really don't feel like it . . . oh, ok . . . if you'll stop nagging." Let's look at a few examples in the spectrum:

- A man nagging a woman for sex
- A man sexually harassing a woman at work
- A man taking a woman on a date and then forcing himself on her

These are all intrusions that women constantly have to deal with. Their genes want a quality man and they do what they can to attract one. But being of stronger size, more desperate for sex, and traditionally violently kept away from women, men have built up aggressive behaviors to take what they can get. Human violence is about men killing other men and raping women. Fortunately, violence is on the decline and technology has created an environment where these behaviors are losing their effectiveness in pursuit of high GPR. I don't assume that we are physically evolving into a gentler species, but with the rise of culture, we are adopting a *nicer* form of violence. Same goal, but fewer people dead and raped.

Struggle (Violence's Little Brother)

Great, physical violence is declining, but struggle is alive and well. Maybe we are not physically fighting each other on a daily basis (yes, some of you are) but we are fighting each other culturally over our status and reputation in the minds of others. To be physically violent, you may have to kill someone, but to be culturally violent, you can demote someone to a lower job function. You can call someone a name. "Dickwad" or "Cunt" (in case you didn't know any names to call people). You can deny someone benefits and you can put down their favorite music. These things lead to cultural death . . . eventually, of course. There might very well be someone at work or at school who is trying to destroy you, and they are doing so by knocking you down in the eyes of others.

It is all very fascinating to watch and the stakes are no less high than when we were on the grasslands of Africa. Kids at school call each other names. Boys separate into nerd and jock camps, the girls into . . . something similar, I suppose. At work, there's always gossip about who's towing the line and who is a waste of company resources. Nowhere is the struggle so clear and so cut-throat as in politics. Two thousand years ago, politicians would have simply had each other killed, but today, as more highly evolved creatures, we might publicly announce the injustice of their tax returns or make up false claims like "He was born in Kenya."

All instances of attacking someone's reputation are attempts to lower the GPP of a person. Physical violence lowers your GPP by killing you or maiming you for life, but cultural violence lowers GPP by lowering your standing in the eyes of the rest of the world. Remember that social standing is a resource just as physical ability is a resource. Lower a resource and you lower GPP and thereby lower that person's ability to proliferate his genes. "Well done," says Nature.

When threatened physically, you go into survival mode — you fight for your life and you do what you have to to get through it. Society accepts that people may defend themselves against violence with violence. But when someone attacks you culturally, they can claim verbally or orally, in public or in private, that everything about you — your thoughts or your identity — is bad and should be expelled. "Bob is a jerk." "Sally is a snake." "Peter has lost his mind." This kind of cultural attack can be just as harmful as a physical attack on your body.

Through culture, your physical being has been spared. You won't be killed as readily as years past; the technology of the legal system has seen to this. But, as we've seen, our leap from biological evolution to cultural evolution has not stopped our genes from competing. Where they once used physical technologies like swords and fists to fight, now they use words and paintings. **Every cultural element: a book, a movie, a song, a beer commercial, communication in a group, promotions and demotions at work are all forms of cultural struggle.** There is no neutrality in these efforts. Even something as seemingly harmless as *Green Eggs and Ham* is taking sides[38]! The genes involved expect results from their investments and they are all driven by the same goal.

Drugs

> ***"Jimmy left for the weekend and now there's no drugs in my room! How am I supposed to have a good time?"***
> — some guy I used to know

[38] If you've never heard of Green Eggs and Ham: http://www.quanology.org/green-eggs-and-ham

Drugs are fun. Is this a surprise to anyone? It's like saying "beds are comfortable." That's the whole point of beds. If you ever encounter an uncomfortable bed, you can rightfully call it a "bad bed." Drugs are a product of technology and technology is here to give us what we want. We want fun. Drugs fill this demand. But genes don't want fun, and therein lies a conflict of interest. We already talked about drugs, and those of you who are high right now are probably slightly stunned by this fact, but I want to show how much of the business world is actually made up of a bunch of drug dealers.

You've surely noticed in life that not all of your activities are directed at GPR – directly or indirectly. What about sports, hallucinogens, music, shopping, TV, video games, food, sex? These are all huge parts of humanity, but you can engage in them without having any intention of improving your GPR.

Drugs occupy that space where cultural evolution has left us some room to mess around and do things that don't help us to pass on our genes. We are so far ahead of the game, we can afford to get drunk and fat. Like a fighter that is so dominant over his opponent that he doesn't even have to concentrate to win. This is the Great Loophole, where much of what we do in our daily lives is a big mistake. It is a gap created by cultural evolution with biological evolution playing catch up. Drugs are those things which do not improve our GPP but don't kill us either. Even though our GPP isn't improving, we are still getting rewarded. Our genes, however, are not.

Drugs, by definition, are delicious. We love drugs. In the name of full disclosure, I should tell you that I used to do pot. I will allow you a moment for the shock of this statement to settle. "Such a bright, thoughtful young man. So eloquent in his prose and so persuasive in his arguments . . . a pot smoker?" It's true. I used to smoke marijuana. I used equipment. I would have snacks and comedy movies all lined up for an evening of folly with friends. I loved it. I still love a drink, a good sitcom, a turn at FIFA on the PS4, and copious amounts of snack food.

What I don't love are the consequences of taking drugs. **Drugs, by definition, have consequences**. They are not an investment in GPR. Yes, you can invest in GPR and lose, but at least you are investing. Drugs are a waste. You might interpret this to mean "don't do drugs" and implicitly it does. But there is no real harm in finding a balance in life. Having a glass of wine here, getting stoned there, listening to music when you could be reading a book. These things are designed to be fun. And they are. And if your goals are set and noble and you are working on them consistently, then no big deal. Again, as with all of this book, it's about awareness. It's about knowing when you're taking drugs and when you are investing in your genetic legacy.

But watch out! Addiction lurks around the corner with drug use. They are the sinister underbelly of drugs. The trap you can fall into when playing with *nature's happy switchboard* — a switchboard that was not designed to experience excessive taps on its circuitry. When you get addicted to the heavy stuff, like narcotics, gambling, porn, and alcohol, you can lose the rewards that drew you to the drugs in the first place and adopt behavior patterns that may very well kill you.

Though almost everyone takes drugs in some form, most of us do not have a crippling addiction. Society would not function properly if we did, and therefore the makers of such drugs need to keep you interested but not desperate. Narcotics dealers know there is a shorter life span for their customers, but the other dealers want you hooked for life, which means you need to stay healthy and have enough money to pay for their product. What kind of drugs are we talking about here?

Much of modern industry is selling drugs. There is always a level of interpretation, so bear with me and think only of the products which really do nothing to improve your life. Some toys are good for a child's development, and some are total junk. But we buy the junk too because it stimulates our child's pleasure circuit. Some video games are educational. Some are not. Grand Theft Auto will not help you improve your life. It is a drug and its makers are drug dealers. I'm not vilifying them or saying they shouldn't make that game. I'm saying that the people who buy it and play it are using drugs. Implicitly, that makes Rock Star Games a bunch of drug dealers. I'm sure they don't realize this, are not bad people out to do bad things, and are completely offended by this statement. Just calling 'em like I see 'em.

It's the same for the alcohol industry, the coffee industry, much of the electronics industry, and most of popular media. The case can be made for much of the music industry as well (more on that in the next section). I'm not putting it down, just explaining what it is. I drink alcohol, drink coffee, watch sitcoms, play FIFA, and love electronics. But when I want to raise my GPP, when I want a deeper, more fulfilling life, I have to put these things aside.

Why do corporations sell drugs? For the same reason they sell anything: they want the resources of someone else. Some companies provide you with things that actually increase your GPP and some do not. The only reason the latter is allowed to exist is because of the Great Loophole. Companies want your money, but they can only get your money if you are alive and flourishing. This is why alcohol companies say "drink responsibly" instead of teaching you new drinking games. They really do want you to drink responsibly. Not for your benefit, but for theirs. Not for some social, moral obligation, but for their bottom line. They would rather have you buy two drinks a day for the rest of your life than 10 a day for 10 years and then lose you to Alcoholics Anonymous, at which point you have to stop drinking all together.

The snack food industry has it even better — the fatter you get, the more you crave their foods and you'll never really reach a point where you have to go into rehab and stop eating chips completely. At least, not yet. . . . Society has accepted fat people, but not drunk people, so the food industry can keep selling you their drugs and taking your money — all the while improving the GPP of their owners. The genes are behind the wheel and the Great Loophole makes it all work.

As long as you are conscious of it all, no big deal if you like drugs and want to spend money on them. I love the show *Arrested Development*. Absolutely brilliant. What I like more than *Arrested Development* is watching it with a bowl full of Fritos on my lap and a cold beer in my hand. I would have a hard time justifying the hours I've spent watching every single episode. Weren't they supposed to have a movie? I just found out there is a new fourth season, so excuse me while I stop writing this GPR-increasing book and go watch it on Netflix. I'll see you in like a week. . . .

. . . OK, I'm back. That was awesome. . . . But at some point you take a look at your life and you see that you are wasting it. Drugs are fun, but there's got to be something deeper right? What about art?

Art

I struggled a long time with this chapter. Is art just a drug? Is it just some strange tool that used to serve as an indicator of physical health but somewhere went horribly out of control? Because, what purpose does it serve? This book has been all about evolution and the technologies that we developed to proliferate our genes. So, how does art fit into all this? It's a big piece of life so it needs to fit into this puzzle. And what the hell *is* art? We know this laptop I'm writing on isn't art, but some painting of speckled dots is. But who cares that a painting is art? Is the painting art because it doesn't serve a function? And the laptop does, so it's not?

Let's start with bells and whistles. Art, like sex and drugs, stimulates the pleasure circuit in your brain. Neurologists can look at the activity in your brain and see that the spots lighting up when you hear a good song are the same ones lighting up when you're sexually aroused. You get a "hit" when you hear a good song or read a bit of top-notch poetry. As with other drugs, this may be just an instance of the Great Loophole where a thing that represents something adaptive gets a free ride on Mother Nature's reward system. We like cheesecake because it contains the sugar we found and desperately needed in the Sweet Spot. So, if sugar represents fruit, what is the adaptive function that art represents? Everything in the Great Loophole once came from something that helped us improve our GPP, right? The lust for cheesecake came from our drive to eat sugar and art came from . . . what?

Harvard Psychologist, Steven Pinker believes that art contains patterns of things that we traditionally like. In his book *How the Mind Works*, he explains that humans have distinct environmental preferences. We instinctively want to live in environments that have pleasant temperatures, plentiful water, plants, and animals to eat. Since some animals eat us, we want that environment to be safe too, i.e. with a good amount of cover and escape routes. We avoid bad environments that are too hot, too cold, too dry, too open, or too barren and lifeless. Both types of environments can be recognized, foremost by *looking* at them and judging them as such.

Now, here comes the interesting part. Visual images, paintings or photos, which represent these adaptive environments are *pretty*. Well, we think they are pretty, but not because there is a nice contrast of dark and light, or that we get what the creator is *saying*, but because **if we lived there, we would improve our GPP**. Maybe we prefer clear water to murky water because we can see the fish we want to catch in clear water, not because it's inherently pretty. Because, what is *pretty* to evolution? Pinker says that even abstract paintings have patterns that simulate the cues of fertile, safe environments. Again, it is not the artistic quality we like, but rather the impact on our GPP that these things represent that we find attractive. He makes the same case for sounds and patterns of music. That "crying guitar" in the first minute or so of "Paranoid Android" is gorgeous, but really it is just mimicking the human sound patterns we make when we are sad. Or so, Pinker postulates.

I can certainly imagine that we are naturally attracted to some things and repelled by others. A painting of a pretty landscape might give me a warm feeling simply because it presents a good environment for me to eat and raise kids. So, it goes naturally that we all agree that a landscape painting is pretty. But what about all the stuff we *disagree* on? There are some chart topping songs that people love, but I can't listen to for 5 seconds. Looks like there may be more to art than the GPP boosts it depicts.

We know that, in genetics, if something adds to the potential for survival and reproduction (what we call GPP), then it is passed on to future generations. If it does not add to GPP, it is killed off. But some things don't make a difference. A peacock's tail is a veritable pain in the ass. Have you ever tried to catch a peacock? It's super easy. Why? The tail, man! That thing, though stunning, makes a peacock easy prey. So, the tail must be adding to GPP, or else nature would have selected peacocks with smaller, less "here I am, come and eat me" tails. So, what is that trade off?

Women. Female peacocks love a man with a bright, flowery tail. But *why* do they love the tail? If the tail makes the peacock that much more obvious as prey, any kids they have would be killed more easily too. Apparently, the tail doesn't hinder their survival enough to cancel the whole project. The tail must come from three selective pressures: 1. The tail is a symbol of genetic quality. Ever tried to grow one of those things, with all the colors and patterns? It's really hard. OK? A peacock's genes have to be like perfect to pull it off. If there are flaws in the pattern, the peahens will notice and blow him off. So, peahens are attracted to beautiful tails, as they represent mutation-free genes. The more ostentatious (look it up) the tail and the more precise the pattern, the less likely to be something wrong with his genes. This means a higher GPP for the peacock and the more likely we will be to have baby peacocks with tails just as fabulous as his dad.

2.The second reason to choose a male with a huge, flashy tail is exactly because it *is* such a pain in the backside. Any male who can take care of his business with that gigantic Picasso on his ass *must* be fit. It's like a guy who can kick people's asses with one hand tied behind his back. When a female sees a peacock with that flamboyant tail, all detailed and still flawless after years of living in the jungle, she can only conclude, "If he's still alive and that tail is still in mint condition, he must be the best choice to be the father of my children."

3. The other selective pressure may just be random. There might have been a peacock with a slightly ornate tail and there might have been a peahen who thought that was cool and decided to spend some time with him. Pure chance. They had kids and their resulting kids had two things:

a pretty tail **and** a pretty tail-fetish. Now the peahens of that brood will grow up to like pretty tails just like their momma, and select peacocks with pretty tails. And next thing you know, we have a whole new species of bird where the men have lavish tails and the women scream for them like Elvis.

The point is that art may have followed a similar route. At the risk of sounding sexist, it looks like art was more of a male thing. We'd be blind to not think of it as more of a male thing today too – not because women aren't good at art, but because men get rewarded more for being artists. As we said, all women in the Sweet Spot were mothers whether or not they could sing. Singing is a very typical female art, but it was not going to make the difference in being a mother or not back in the Sweet Spot. But that may have been more the case for men. Look at some of the popular male artists of our time and think: where would they be without their art? Tom Petty? Kurt Cobain? John Lennon? Then contrast that with some females. Adele? Jewel? Beyoncé? Take away their talent and imagine how their GPR would fare.

So, men started painting things on walls and this was recognized by their little tribe and their GPP went up. It doesn't have to be much, remember? Just a little bit over generations starts to become noticeable. And before you know it, just like with the peacocks, the men have an artistic flair and the women really like it. The ensuing children have both: an artistic flair and an appreciation for art.

But art is more than just a something we *like*. **Good art is *hard*.** Maybe once you tried to write a song or make a painting. Didn't it suck? Half the time, our reaction to good art is "*How did he do that?*"; the other half is the more depressing version "*I could never do that.*" Whether you could or couldn't, the point is that not everyone does. Those who create high quality art are somehow in command of their environment and mental, as well as physical, faculties in ways that the rest of us are not. **Good art can only come from the combination of intelligence and confidence**. Art is about making decisions. "Do I put the line here or there?" "Should it be red or dark red?" Decisions in art, and decisions in life in general, take guts to make. That is why confidence can be *felt* in art. Reviews of albums contain words like "bold" or "flimsy" to indicate the level of confidence the listener feels from the artist. Artistic types are often extreme individuals — not like the rest of us who live our lives as a sort of compromise, never making the bold decisions to go one way or another. And good art is polarizing because it makes those bold decisions. It appeals to some people and not others. What it tries not to do is appeal to everyone, because that would mean the exact compromises which signify a lack of confidence.

What about "mainstream music"? Doesn't that appeal to everyone? Yes, that's its job – to appeal to as many people as possible. That appeal also involves decisions, not confident ones, but *calculated* ones. What we often see and, some of us feel, behind the pop music of today is the *committee decision* of the music. The appeal of music is that it is this perfect display of intelligence and balls. The committee takes away the balls and replaces personal intelligence with group intelligence. The group (record companies) will still try to present the music as authentic, because they know that is what the people want — we are drawn to the expression of an individual. What they keep a secret is that a professional song writer wrote the song (not the singer), the board approved it, and it was given to a producer who is rewarded for how many copies it sells.

But I digress. . . . Nature has selected good artists partly because they display the intelligence and confidence that puts them on the top in a group. If those qualities are not always present in some famous artists of today, the drive to revere artists is still there — just like the white tail signal in the zoo.

We will take the appeal of art a step further. We like art for more than just what it says about the creator. We are attracted to pretty things, but not all art is pretty. One song that pops off the top of my head is "Kim" by Eminem. It's amazing, but before you go queue it up, fair warning: it's an acquired taste. Is it good musically? I'm not sure. Are the lyrics good? They're OK, I guess. But the story and the way it's told is unbelievable.

Is it original or unique or interesting? No, not really. It's the *emotion* that gets me. The music, lyrics, and delivery create an insane moment where a jealous guy kills off his girlfriends' lover and is now about to kill her. I'm right there and I feel what he feels. At that moment, he's lost it. All the pain in his life has been building up and the dam has broken. Eminem lets it all out in this song, uninhibitedly showing us both his anger and his defeat. There is no hesitation or doubt. He sounds drunk, further showing how much he can't deal and doesn't care. I am seeing his deep emotional innards and I am sucked in — forced to either love him or hate him.

Art often has a story and that story brings us into a situation where we experience the same emotions as the people involved in the story — emotions we miss in our lives simply by the fact that our lives are not constantly filled with such drama. That's the whole point of movies or songs — to show situations that are remarkable. If it were normal, it wouldn't be in the movie. We like interesting emotions; good ones feel good and the bad ones also feel good in the sense that they *teach* us. As Pinker explains, extreme emotions mean that things are happening. Fortunes are being won and lost. Kings crowned and decapitated. Lovers adored, ignored, united, betrayed — some even killed. All these things deal with swings in GPP — people go up in life or they go down. If the story shows a jump in GPP, we can enjoy the sensation *we* would feel if it happened to us. If it lowers GPP, the story can serve as a warning to us to not fall into this trap. Music, movies, and literature are the popular art of our day — not paintings and dance. In this sense we can see that it is because these things best convey stories, because they use words and real images. That is why we don't care for foreign language music as much — we miss the stories. There are some awesome Russian bands, but if they sing in Russian, I've lost something. Language is the key tool which conveys complex situations — something that paintings, dance, and pantomime have a hard time keeping up with. We love pretty sounds and we love pretty colors and patterns, but the real sell in art comes from the stories.

What those stories do is get us on the same page. The more personal a work of art is, the more the message can be seen as an expression of personality. You don't so much connect with a message, but with its creator. **If emotion is the glue of learning in our brains, then art is the glue of learning *across* brains.**

When you are exposed to the thoughts and ideas and "personal expressions" of another human being, you learn to empathize. Art stimulates empathy since art is a reflection of the emotions, thoughts, and ideas of someone else. When we understand that other people feel the same as we do, we are more apt to adopt the Golden Rule. "She likes pretty waterfalls just like I do." In that subconscious thought, you become equals with that person and thereby move in the direction of appreciating his wants and drives as much as your own. **In cold evolutionary terms, that empathy you receive as an artist puts others on your side**. Not only are others now more willing to help you, they are less willing to hurt you. Violence against others, injustice against others follows naturally when we think that the other person is a sack of dirt who deserves it. Art cures this thinking. I love meat, but if I read poetry written by cows, I would turn vegetarian overnight.

Quick personal story: our dog just died. 13 years old, she was. It was heartbreaking. If you have an old dog, you know that most of them don't die naturally, you have to put them down. She was limping on three legs and in enough pain that we had to make the decision to put her to sleep. We made an appointment with the vet on a Friday afternoon. And there you are, it's Friday afternoon. She takes one last pee and then walks to go back to the house, but we don't let her go back to the house, we pick her up and put her in the car. The kids wanted to be there too. It was gut wrenching. Everybody cried but me. I just didn't cry. Not the crying type. The next day we buried her at a pet cemetery. More crying, but not by me. As we left the cemetery, the guy there handed me an envelope. "Read this tonight; it will help put everything in its place." It was a poem. We went home and a little later in the day, my wife opened the envelope and started reading. Two sentences into it I started bawling.

The poem was read from the perspective of a tired old dog at the end of the line. He was telling his owner it was OK to let him go — something that all of us had an awful time doing. I'm fighting back the tears thinking of it now. It put into words what no dog can say but what we can imagine they feel. It was through those words that I could finally be connected to the whole situation and feel the heaviness of what just went down. It pulled me out of my emotional routine and pushed me full-force into a connection with a dying dog.

If art tries to put us on the same page — be it the page of some movement, group, or concept or just the page in some guy or girl's head — then what it does is move us *forward.* Every time you take on a new perspective or solidify an existing one, you move forward. It may be a bad move, but don't worry: nature soon kills it off. So, again, we don't need to be the ones who judge what "forward" is; nature does that for us. A song might have a racist message or a loving message. When you take on either as a new perspective, you are moving forward. Now, racist songs aren't fairing very well, which tells us what nature thinks of all that, while songs of love are topping the charts. Think of it as a new adaptation on an animal. It is *new or it is stronger,* so it is *forward.* But nature decides if it is good or not. The more adaptations, the more progress. Nature sorts them all out, we just have to provide her with material to judge.

In this sense, art is a catalyst for progress. If we go into our gene-detective role, we should be cautious to think that genes would spend so much time and effort doing something that has no benefit. Through this mindset we can see art working to advance the genes of the individual and of the group. Yes, it has drug-like qualities; you can get addicted to it. I'm addicted to Green Day at the moment. I'm not sure I see how they are helping me to progress culturally, but I just love their songs and their vibe. It's fine if the progress is all subconscious. But it seems that art originated from a happy place and has not only served man-kind well, but has given us a meaningful connection to others that, by itself, feels like enough to make life worth living.

Charity

We've talked about some mean stuff in this book. Some dark sides to humanity whose very drive has been etched into our brains by natural selection. "Apparently, nature wants us to be mean." But no, we know better than to conclude this. She wants us to be successful. In the beginning of this book, I said that murder and rape were natural, but so was kindness to complete strangers. Here I will finally make good on that statement and explain how we are also naturally nice to people. You can probably guess that it involves evolution.

Shout this, would you? *What do the genes want!?* "Proliferation!" *When do they want it!?* "Now!" Awesome, thanks. One thing is for sure: any happy, sweet, friendly drives in humans, in any animal for that matter, is there to simply serve the gene-ish drive for proliferation. We like to give, don't we? "'Tis better to give than to receive." So true. I love to give people things . . . especially if they appreciate it. Ah, now we see the Gene-ish Self. So how does giving things away help us? There are several valid answers and we'll go into the obvious one first.

Obviously, we give people stuff so that they will someday return the favor. "Tit for tat." Chimps do this on a very complex level. They give to one chimp and expect something back, and if they don't get something back, they get mad. We give things to people and we might say we don't want anything back, but we want *something*. At the very least, we want that person to think better of us than they did before. We've had a good look at how people team up in the previous chapters and we know that your social network is a powerful resource in your quest for high GPR. Having other people who "owe you one" is doing nothing other than transferring the resource you had in one category to a resource in the social network category. You give 5 people an apple each, and instead of having the resource of 5 apples, you now have the resource of 5 people who owe you something. It's the same as taking $500,000 out of your bank account and buying a house with it. You haven't lost the resource, you have converted it into another form.

Now, in the Sweet Spot there were no houses, so giving things to people was a great way to convert surplus resources. If you can't eat your 10 apples, you might as well give 5 away before they rot. Natural selection found this to be a valid strategy for gene proliferation, and thus, the drive to give extra resources to others was built into the human psyche.

On the other side, the drive to reciprocate is also built into the human psyche. Whenever someone gives to us, we feel compelled to give something back. If the flipside wasn't there, natural selection would have found giving to be a dead end. They go together. I just bought a new car (well, used, but it's new to me), and the dealers *gave* me a good price for the trade-in. But it wasn't a good price — it was the same price that other dealers were offering. But because it *looked* like they now gave me something, I felt as if I owed them. So, I didn't haggle on the price. This is a common practice among sales people: to make it look like they give you something, when actually it's something that costs them nothing and something you didn't really care about. But your piece-of-crap brain still thinks you owe them something, so you make a concession too, just to match theirs. So, the drive to give is strong, but it is matched by a strong drive to give back. And that's what we want, lots of people who want to pay us back. The more people we have in the world who owe us, the higher our GPP should be.

We not only want people owing us, we want people thinking well of us. There are many kinds of people in this world, but our brain takes the easy road and categorizes. "There are two kinds of people in this world . . . " and you can make up whatever two polar opposites you like. I've heard "jerks and chickens," "good girls and sluts," "wimps and bulldogs," and I've even made up "crazy and repressed." In terms of exchange, you have "givers and takers." I will ask you now to think of some people in your life and see if you can classify them as givers or takers. Go ahead, take a minute. . . .

As I flip through the contacts list in my head, I notice that many people are actually hard to place, but some are glaringly obvious. "This person is a huge giver," and "That person is a definite taker." We hate takers and we love givers. In being a giver, you gain more than just those things that people give you in return — you get a good reputation. You are deemed trustworthy for future exchanges with friends of friends. Remember, it's not who you know, it's who you know and what they think of you. If they think you are a giver, you are much more likely to benefit from that relationship.

Call it cynical, but that is exactly what we want to do with every relationship: benefit. That doesn't mean we are heartless, conniving bastards, it just means we want pleasant relationships. We don't want people to be abusive, take advantage of us, put us down, or in any other way provide a bad experience in our interaction. We want nice relationships. **A "nice relationship" means that we benefit somehow**. That benefit may be an emotional benefit, it may be a physical one, but it may just be a mental feeling that you and the person across from you are enjoying each other's company. But in all of these cases, there is a benefit, which is recognized by the genes as an improvement in your GPP, and they provide you with a warm, fuzzy feeling that you are doing well. Can you think of a beneficial relationship which does not somehow improve your GPP? I'm stumped.

So, we're not cold and calculating, but our subconscious does recognize a few things in the person we are giving to. Things like, their potential to give to you someday, their GPP in general, and their level of need. Let's have a look at each.

The person you are giving to is only worth giving to if they have the prospect of giving back. If they have nothing, you may still give to them, as long as they look like they may have something in the future. We don't readily give to homeless people on the street. Why not? Because, we know what they will look like in a year's time. But giving to, say, a little kid is different. He may have nothing now to repay you with, but he'll grow up and remember that time you gave him a soccer ball. My boys still have the soccer balls Uncle Tom gave to them years ago. One day, he'll collect big time.

Next, perhaps counter to our intuition, **we give to people who already have stuff**. People with high GPP get free stuff more than the rest of us. They may be rich, so you could classify them in the first example as someone who could easily pay us back, but it's more than that. We give to high GPP people because we want to be associated with them. "Cool by association," as it were. The more we give to high GPP people, the closer we get to them, the more they like us, and the more the rest of the world sees that.

Sometimes we give to people when they directly ask for help. Our subconscious knows to place value on the level of their need. They say the best way to get money is to act like you don't need it. But that relates to the previous example of looking like you have high GPP. If you are desperate for something, someone may very well give it to you because you will now "owe him big." This means giving small things to desperate people can still offer a good return, even if the person does not have a lot of means to return the favor.

In all these cases, one thing is clear: The person to whom you are giving needs to have some level of worth. A world where some people have lots and others have nothing is not good for business. Sure, we want to give to people who already have enough, but in general, coming regularly in contact with people who have nothing does not sit well with our genes. Not because we have a moral obligation to take care of them, but because we don't want the hounds at the gate. We want to make sure people are content enough with what they have so they don't come take our stuff. Because when they do, you know they're gonna take *all* of it.

Appeasement is another motive for charity. I don't know about you, but I find it very unsettling to think about places in the world with no government, mass killings, mass raping, kids carrying guns and sleeping in the dirt. "We can solve that for 10 bucks a month? Let me get my check book." I know we can't really solve it for 10 bucks a month and you knew I didn't actually have a check book, but the drive is there and we're moving closer to a place in time where we can correct these situations through organized giving. That doesn't necessarily mean money. We can give them something more important: technology.

We take a look now at family, those people you buy Christmas presents for. From a gene-ish point of view, it is obvious that family should be the individuals who you give to and receive from most: they share most of your genes. The genes see their brethren in someone else and are quick to share whatever they have left over. Case closed. But wait! Since the dawn of cultural evolution, we see that we also have a cultural family. Those people you share ideas and beliefs with. We said that part of your cultural technology is getting people to ally with you and this bit on charity shows how we more easily give to those with the same interests and viewpoints as us. We give to charities that we *believe* in. We support projects and initiatives that are an expression of our internal culture. I'm not giving to an Islamic Community Center any more than Ted Nugent is giving to the Socialist Party. But I'll give to a foundation that supports clean environments and Ted will give to an organization that puts a hand gun in the desk of every fourth grader in America.[39] Your cultural allies are also family and you give to them, too.

Let's now move from the individual to the group point-of-view — that should be easy for your experienced self by now. We're more than half way through this mother, so I'll assume you've got the swing of it. Groups also give and expect a return. But when a group gives, we can take out the pleasant, emotional aspect. For instance, when America gives $5 billion in aid to Rwanda, it does not walk away with a proud feeling of inflated worth. As opposed to individuals with their faulty conscious interpretations of their actions, groups know exactly what they are doing and why. A country's foreign policy is based in "the carrot and stick" approach. Charity is the carrot. **Countries always expect something in return for giving, and often that repayment is inked at the moment of issue.**

As with everything else in this book, when trying to understand a phenomenon that covers the globe, we ask, "How do the genes benefit?" Genes that give up their valuable resources expect something

---

[39] I'm just bustin' your chops, Ted. Of course, fifth graders too.

back for it. I personally don't like the idea of blindly donating money to some cause — not because I don't want things to be better, but because I don't think that's the best way to do it. But many of us regularly donate money to organizations. We can think we don't want anything in return, but maybe, at the very least, we want the freedom from the guilt of not giving.

Not to get too cynical, but when we give money to poor people and poor countries, we deprive them of our greatest resource: the technology to create their own wealth. That is the valuable resource we have and they do not. Technology is why we are rich and lack of technology is why they are poor. This applies to homeless people and to poor countries. Yes, they need money to apply the technology, but giving them the money without the technology to invest it is a recipe for waste.

More than ever, we live in a society where everybody is giving to everybody else. Early in our evolutionary journey those we gave to were limited to our own tribe. Through technologies such as trade and cultural groups like governments, religions, fan clubs, foundations, and various causes, we are giving more and more to people we don't know but whose interest is tied to ours. We have never been so dependent on so many people as we are today. Our fates are held in the hands of people all over the world, and we want to see them do well so that *we* will do well. This is built into the fabric of nature without anything supernatural or moral. Personally, it helps me sleep better at night knowing this system is in place and is created by something real and robust, something so elemental to our world. It has taken millions of years to develop, and we are so lucky to be living on the tail end of it.

What do the genes want? They want to proliferate. That means everything needs to be working optimally in the world. If one group dominates, it means that others suffer. At first glance this sounds like a good thing for genes, but if we understand that by limiting our group, we lower the technology and we lower the number of people we can benefit from, we see that this is, in fact, not a good strategy for our genes. Our genes are best served by a world where the people in it are high quality, where the possibilities to innovate are allowed to blossom, and where everyone chooses non-zero sum competition over war.

Progress

> ***"We got out of the food chain. That is a massive upgrade."***
> – Louis C.K.

One question we all ask ourselves at some point in our lives is "Would we be happier living at some other point in history?" Many people fantasize about the past and how it would be to live 100, 200, or even 1000 years ago. Maybe you've thought about going back even further. We're not sure if it would be better to live back in time, but we consider it, and often romanticize about it when we're feeling fed up with this life.

Well, we can't go back in time, so the question then becomes: Would we even want to? Is life better now or is it worse? Or do things fluctuate throughout history regardless of the arrow of time? **To understand our progress, we have to understand where we've been**. Only when we can place ourselves in the lifestyles of history can we even start to consider the present. In order to do that, we need an objective look at how things are now — not some emotional outcry like "I don't want to live in the modern world."[40] The drudgery of my middle class life can often leave me thinking, "This place sucks, let's move to the year 1621," but that's just my emotions talking — I'm not really thinking it through.

The modern world is teaming with war, rape, murder, violence, and slavery. Even a little bit is too much for us, and we criticize the slightest instance of any of these things. But the statistics show that all these things

---

[40] Name that tune!

have decreased if you factor in the number of people on the planet[41]. And that's a big factor to weigh in. There are reports that there are more people in slavery today than there were at the time of the Emancipation Proclamation. While that may be true and awful, what we fail to mention is that there are five times more people on the planet now than there were in 1863. If you see the amount of people in slavery compared to the amount of people on the planet, the percentage of people in slavery is far lower than ever. I'm not saying we can shut down the charities just yet, but I'm saying there is a positive trend away from slavery. We're getting there.

These are the sort of objective arguments that our strung-out emotional brains hate. Those arguments say that our dissatisfaction is unjustified. Someone was complaining to me today about how boring her job is. I get it, man. I once had a boring job and I went crazy. Putting someone behind a desk and telling them to do nothing for 8 hours is a form of torture. But it's still progress if you compare it to working 18 hours a day in a sweat shop or on a farm — which is what she would have been doing 100 years ago. Most of the time, when we dump on the modern world, it's because we're having a hard time with it or we just saw some news about something horrible.

And is there anything else on the news? No, not much else. The other problem that optimism faces is that we are bombarded with negative images. Yes, awful things are happening. No doubt. But awful things were *always* happening, we just never had it shoved in our faces all day long before.

There are reasons to be optimistic today, but there are also reasons to be pessimistic. We are bound by our psychology, and our brains are not always the best at rational thinking. For example, your job might be to solve one of society's problems using technology. When it's your job to solve a problem, you subconsciously tend to make that problem out to be worse than it is. As we are judged on our technological contribution

---

41 Have a look at *The Better Angels of Our Nature* from Steven Pinker for details.

to the world, it is natural that we try to make the problem we're solving seem bigger than it is. But serious conflicts arise when a person is told to work hard to solve problems and then realizes that those problems are not going to kill him and everyone he knows. It is human nature to think in catastrophic terms.

Am I saying that all our worries about the world are exaggerated? No. Have I gone from a cynical person to a happy, optimistic one just by being objective? No. Just saying that there is a lot of unfounded negativity going on and it mostly stems from the news, our misunderstanding of the numbers, and the psychological tendency to make things out to be worse than they are. There is still much work to be done, still a lot of close-calls to come, and a lot of forests that will be paved over and oil tankers that will crap in the ocean before we have a world that we unanimously feel great about.

But let's take a really objective look at what's going on. Imagine for a minute that you don't have a meaningless job and the world is not out to get you. In fact, let's take a step back from your life completely and think about the world in general. Remember where we have been and compare it to where we are now. For me to list statistics would be boring for you and a lot of extra work for me, so let's skip it. Up until this point, the picture I've painted of our ancestors comes close to the English philosopher Thomas Hobbes' image of a "poor, nasty, brutish, and short" life of an individual. Anthropological studies seem to support this. High murder rates, incessant war, constant illness, unequal societies, rampant slavery, poverty, inequality for women, horrible cell phone reception, and even higher taxes than we have now. But at least there were no commercials. . . .

Society is in no way perfect, but at least we can get excited about the fact that we're moving in the right direction. Think about these trends and we can see that even if we are not in heaven, we may be in the car ride on the way there. In no particular order. . . .

- Increased democracy
- Better treatment of children

- Better healthcare
- More comfort and luxury
- More hours of leisure
- More fun activities
- More access to mates
- Reduced violence
- Better and cheaper access to information
- More access to healthy, varied foods
- Better, faster, cheaper transportation
- Better, cheaper communication
- More transparency in the work of individuals and organizations
- Dispersion of power from fewer to more people
- More self-made individuals
- Better rights for women
- Better rights for minorities
- More effective justice system
- Greater interdependence between people and countries

None of these things are perfect, they are just better than ever before. It's notable that all these things are a result of technology, not "a more finely-tuned moral compass" or even good decisions by our leaders. To detail each one of these would take a whole other book. Fortunately, you can pick up a copy of Matt Ridley's *The Rational Optimist*, in which you will find an in-depth look at many of these trends[42].

Through natural selection, we have arrived at a place that is very good for our genes. "Never better," so to speak, and tomorrow will be the same. The good part is that what's good for our genes is good for us. Their success is our fulfillment. All the cultural progress we see as a result of technology, also results in successful genes, and we share in that triumph. We are fulfilled, happy, or whatever you want to call it when we

---

[42] Illuminating look at how progress is built into the fabric of human evolution: http://www.quanology.org/matt-ridleys-the-rational-optimist

are able to support our families well, and when we are able to contribute to the progress of our society. Hard work and competition will never go away just like the 2nd Law of Thermodynamics will never go away, so you can stop waiting for society to "get it." We can, however, have an appreciation for a system that puts us to work toward a silly goal, but at least rewards our brains occasionally in the short-term and takes up our list of grievances in the long-term.

Of course technology is a double edged sword. **You can't have technology without someone using it to do bad things**. We can still be grateful for an "auto-correct" feature that eventually finds the abuse and develops new technology to weed it out. Someone invents guns to kill enemies and defend himself, then people use guns to kill innocent bystanders. New technology is then invented to assess a person's legitimate right to own a gun and laws are made to enforce the technology of that sound gun-owning policy. America may even adopt these technologies.

Yes, there are still bad situations. The rich are getting richer and the poor, poorer in some places in the world. But we all agree that we don't want it this way. Even many rich people agree. So, what can we do? As we've seen, natural selection takes care of these things. Doesn't mean we don't have to do anything, it just means we can't not do anything. You can rest assured that it will all happen. It's just a question of time. And guess what speeds things up. . . . [43]

Technology results in a higher quality of life for the most people possible. There will always be a priority list — people in first-world countries will enjoy the benefit of technology before the people in third-world countries. In fact, that is basically the distinction between first and third-world countries: the level of technology that they hold. But even in North America and Europe, there are poor people with a poor quality of life. They are less than happy in this modern world and feel as if the system is not working. It is working, but perhaps too slowly for their tastes. Even the poor people are better off now than they once were.

---

[43] Technology, in case you were stumped.

They have not yet gone from poor people living in dirt, like they did 100 years ago, to living in free-standing houses with landscaped gardens, but conditions have improved. There is still a high murder rate in some bad neighborhoods of America, but it's not *20% of all males.* Though 20% of the males might be going to jail, they are not being murdered[44]. That's evolution — it's still slow, but the trend is there and we can speed it up by recognizing it and consciously supporting it.

More technology? Yes, that is the answer. Before we get too scared of robots taking over the world, let's remember that *we* are the real impediments to technology. Technology must make money and we are the ones who need to pay for it. If we are not ready for it, it will not happen. There are many technologies which are possible at the moment, but are "ahead of their time" which is just a way of saying, "People won't pay that price yet." And remember that technology is not just circuits and plastic parts. It's anything that improves our lives.

The crazy part is that this is all driven by genes. **Genes who want more become people who want more**. Those people develop technology to get more. And those people must consider other people when executing their technology. The world becomes an orgy of inclusion. For me to have a high quality of life, I need your help and that means I need to respect your drive for a high quality of life. I may have to compromise on what I want, but it beats spending the energy to kill you first in the hope that I can still get all the things I want without having the things that *you* used to produce. And you are producing a lot these days. Thanks for that. It would be a bad move to kill you.

In fact, so many people are producing so much, and all of it seems to find a way into my life. I have a laptop made in Asia, I just had some fresh squeezed orange juice from African oranges, I'm staring at a bottle of Spanish wine, etc. Literally thousands of people from all over the world were involved in creating all the junk in my house. I wish them

---

[44] For the record, I feel that America's prison system is one of the worst things about America.

well, for when they are well, I get more junk. And they wish me well, so that I can pay them. Increasingly, the quality of our lives rests in the hands of more and more people today.

Morality

In truth, there are a lot of bad apples out there. I want to love people. . . but they make it so hard sometimes. Are people good by nature? Bad by nature? What's the state of morality in today's world? And can someone with an evolutionist point of view even talk about morality?

Comedian Steve Harvey said on *The Tyra Banks Show* that atheists have no moral compass. This is indeed a big argument of religious groups — that we need them or the whole world will go to pot. Morality is about right and wrong. Maybe you've noticed through the course of this book that right and wrong are not appropriate terms when talking about life — even human life. Just like blue is not an appropriate term when talking about the stock market.

Male lions kill the babies of rival males. Male orangutans rape females. Chimpanzees fight wars. Black widow spiders eat the males who just boned them. Frogs eat each other. Ants enslave other ants. In all of this, there is no morality to speak of. None of it is *wrong*, because all of it is natural. But "Humans should know better." Huh?

Many people hold on to the idea that without morality, we would all be killing each other's babies. Hardly — our goal is genetic proliferation, not violence. Violence has only ever been a means to an end. And, yes, there has been much violence in the course of human civilization, but that's only because it worked. There was not enough technology in place to stop it. That's changed quite a bit. I mean, it still works, but less so. My religion teachers used to argue that morality is not some agreement we all simply made with each other — like "Don't kill me and I won't kill you and we can then both relax a bit." But that's exactly what it is. The only twist is that it's an agreement we made subconsciously through natural selection. For ages, the tyrannical, the blood-thirsty, the despotic, the murderous, imperialist, bellicose, and dominant ruled the earth. Despite that, the people who found ways to put their energy to use in industry and cooperation were rewarded through prosperity. Eventually that circle of peace and honesty expanded to other groups. We call this morality, but all it is is a strategy for improving GPR. Look at any moral code and the outcome always leads to prosperity.

The problem with moral codes is that they are general. Not every situation can easily be judged as beneficial or harmful to the group in the short-term or long-term. Strategies change over time. Rape is not prohibited by the 10 Commandments. Maybe they thought it was an acceptable strategy 3500 years ago. In fact, it is a strategy, but not a very good one, and it's getting worse all the time. What's changed? Power has flowed into the hands of women who have grouped together to condemn it. Men, through greater connectedness, better realize that not condemning rape globally implies condoning it locally — and that could be your sister. "Privacy is dead" noted some rich guy from Boston — and that means that forcing yourself on that drunk girl at a party will be attached to you for the rest of your life (it just happened to a bunch of teenagers in Ohio and now their names will be linked to that incident for the rest of their lives. I'm Googling it now.). . . . It's not OK, you know it, and society gets better for you behaving yourself. And ultimately, you're much better off too.

Technology has taken us to that better place. I could not have found those names that easily if it weren't for technology. They wouldn't have been caught if it weren't for technology. Technology will see to it that this crime follows them for the rest of their lives wherever they go. The story with our world is that people are getting what they want — which means they are also avoiding what they don't want. We don't want to be beat up, we don't want to be raped, and we don't want to be lied to, so through technology our society slowly moves in the direction of peace, respect, and truth. I would argue that writing these rules down and putting them in law books is only a measure to hasten their adoption — they would have been applied to society in another form any way. It's just good business.

A world ruled by natural selection is not a perfect world, but it is heading in the direction of perfection by itself. Our subconscious desires are the judge of everything that happens in our world, and those desires were created by evolution. We know what we want, and we respect everyone else's right to go about getting what they want as long as it does not stop us in our own pursuits. We implement technology to make it all happen. That's really all the morality we need.

Therefore, to talk about people as "good" or "bad" is not suitable. We can expect people to do what works — it's in their nature to do so, just like it's in an apple's nature to fall to the ground. There are no bad people, just people I would rather not deal with. And me not wanting to deal with them should be a disadvantage to them. If enough people don't want to deal with them, we exile them from society. Hell, I want the whole world to be involved in my life, and that means I need to do positive things for society. The more we reward people for contributing to our lives, the more we support the strategy of a traditionally "good person." And we all want people to contribute to the quality of our lives, right? I want people to build good roads, good schools, farm good food, write good television shows, etc. And I will pay them good money to do just that.

We don't have to wait for God to judge people who act like jerks. Right now on this earth, crime is its own punishment in that the bad apples live a life of less value because of their negative behavior and attitude. We still have a lot to teach them about being more amicable, but subconsciously we are already doing so by raising global awareness about the personal rewards of doing positive things. Slowly but surely the environment supports jerks less and less.

Some junkies were toking up at the train station in the little shelter the other day. I told them, "You can't smoke here." I'm such an old fart, but I'm thinking, "Think of someone else for a change." More and more, society is pushing people to do just that. It's only the outcasts, the addicted, the unemployable who are not on the same page with the rest of us. We all have to be considerate of other people or else we will fail in life.

Anti-social behavior involves more than those dudes in the corner of the party who won't talk to anyone, it means not getting on board with mutual respect and consideration in society. Fortunately, it's being weeded out by natural selection like short giraffes. **Morality is being created not by God, but by desire – the very thing they told us to watch out for**. The more we want, the more technology we have to utilize to get it, and that means following rules for working together.

Finally, we realize that no one has good or bad intentions, since we all have the same intention. It's just that some people have a friendlier, inclusive strategy, and others have a more malicious strategy. But we all want the same thing. We have bumper stickers that say "Mean people suck" and indeed they do, but we cannot will them away. We have to have faith in the system (natural selection) that it will all work out. We are in a sense, in the hands of a god, but a very unconscious one.

Digital Evolution

We are evolving digitally and we don't even know it. The same thing happened with cultural evolution — its presence was so gradual that we didn't even realize it had started. We don't really notice digital evolution until someone with way too much free time, like me, notices it, sticks a label on it and then shows you all those things in your daily life which are actual evidence of digital evolution. And then you're like, "Oh yeah, of course. I can see that."

I've told you that technology is not just about sensors and switches, but this part is. We are flesh and blood and as such are limited in the physical actions we can take to achieve our goal of gene proliferation. Then along came cultural evolution and opened up a whole new world of possibilities where, not only could we improve our resources through technology, but our thoughts and ideas could cross over into the minds of others and thereby extend our reach and physical influence on the world, even after we died. Digital evolution continues in this vein by taking our thoughts and ideas and translating them into computer code so that computers can improve, augment, and spread them on a level we could never have imagined even 50 years ago. This may sound like science fiction, and it curls my lip too, but we're already knee deep in it.

In cultural evolution, our genes started using the strategy of working together, learning from each other, and inventing technology. The people who were too cool to do these things died and took their technology-averse genes with them. In digital evolution, we see our genes adopting a new strategy — digitizing their thoughts, ideas, knowledge, and very presence to allow computers to process them. Sounds weird but if you posted on Facebook, YouTube, or Twitter today, that's exactly what you were doing. And it was your genes that subconsciously guided you to tweet though consciously it was the ridiculous video of the guy with the dead-cat-copter.

What's the worst thing you can call a human being? Take note, it's not he's "ugly" or "weak." Some fish and birds would be destroyed if you called them ugly. An elephant would slip into depression if you called him weak. But humans? Our zinger is "stupid." Remember *A Fish Called Wanda*? Otto kept saying, "Don't call me stupid."[45] That's the last thing we want people to think of us, and we go to great lengths to cover up how stupid we may actually be.

Intelligence is a big deal; it was a major component in our rise to world domination. Knowing things has always given us a leg up in our competition with other people. We only need to be so smart to compete against other species, but we need to be exceptionally smart to compete against other people. Knowing where water is, where food is, how to make tools, how to make weapons, how to get other people on your team, which berries will kill you, etc., etc. This is the sort of knowledge that made fathers out of ordinary men, and the intelligence needed to know these things got successfully passed on. All of culture is based on

[45] What a fantastic movie: http://www.quanology.org/otto-says-dont-call-me-stupid

information. And intelligence is not only about collecting information, but processing it. We need to know specifically where water is, and also be smart enough to know where water *might* be if placed in a new environment. Most of us are actually living in a new environment. We migrated off the African Savannah some 50,000 years ago and ever since we've been tourists in a strange land. "Water *might* be near where plants grow." "It *might* be at the lowest ground I can find." We take in the information around us, process it, and then come to conclusions.

In today's world, we are awash in information and there are so many opportunities to use that information for our own benefit. The problem is that there is only so much information we can remember. Many of us are plagued by recurring nightmares, not about being killed or terrorized, but about forgetting important stuff. Even if we could remember it all, we're not always smart enough to know what it all means or make good decisions based on all the information we have. We've always needed help. Just like we had tools to do physical things, we have mental tools to help our memory and aid in our decision making, such as paper to write things down and maps and charts to see things in a visual way so that we don't have to spend so much energy "seeing" them in our heads. As with all tools, the people who used them benefitted and improved their GPP. The tools themselves evolved and got better at helping us, so much so that we reached a point that we didn't need to evolve biologically anymore. But evolution continued, just in a new form: culture. And then along came computers. . . .

Computers are the newest tools — we had the Stone Age, the Bronze Age, and this is the computer age. We would be blind not to understand that computers are evolving, just like their stone and bronze grandfathers. We started with hard drives the size of refrigerators and have now put the same amount of power into a chip the size of your thumbnail. Meanwhile, there is a tortoise somewhere who will soon turn 100 and hasn't evolved his DNA one bit (no pun intended, but I'll take it). Every other aspect of computing has followed a similar path: graphics, speed, processing power, software, connectivity, and even . . . ease of use.

What Are Computers *Really* Doing?

Yes, computers are a pain in the ass. Coincidentally, mine just crashed five minutes ago, and I was worried I lost that last paragraph (I didn't!). The truth is, I could still write this book exactly as it is without a computer, but it would take much, much longer. We are where we are, not because any one of us is so brilliant, but because technology has allowed us to store and share the "how-to" of every little advancement we've ever made. "Standing on the shoulders of giants," remember? But we're not standing on their shoulders, we're just learning everything they know because they wrote it down. Writing things down proved to be a way to hold more complex instructions.

Try to think of a good joke.... Go on.... You've heard many in your life, but I'll bet you can only remember one or two right now. Thankfully, all the jokes ever told are stored on computers and if you take a five minute break and go search for some, you will be overwhelmed with the results. Don't bust a gut laughing. Moving from written language to computer language has allowed us to store and process more human information than ever before. You may be thinking, "John, my computer understands human language. I write in it and I can even talk to it." True, but there is always a program to translate it for the computer. Your computer doesn't store the word "the" as "the"; it stores it as "01110100 01101000 01100101". That's what I mean by computer language. And when it's in this form, it is digital, and it can make use of the power of computing.

So how do computers process information?

- By algorithms – as in: what search engines use to determine the order of search results. Algorithms are nothing but rules — "do this when that." But they can be super complex when they are in computers. I can only go through a few rules at a time in my head but Google takes billions of web pages and applies hundreds of rules to them and then re-sorts and filters them and presents them to you in less than a second. Holy shit!
- By interpreting — merging two bits of data into a meaningful new bit of data. As in: knowing that you are 40 and male and then sending you a brochure for motorcycles.

- Allowing us to propagate the information to other systems and ultimately other people. As in Facebook: you upload a picture of your little girl at ballet class and now everyone you know can see it.

Remember that we are machines built by our genes to help those genes replicate themselves. And remember that we developed thoughts, information, and ideas as a way of linking up people in a working relationship to cooperatively replicate genes. Computers are a tool to help do this better. Not only can we inject our thoughts and ideas into other people, we can inject them into computers who will then inject them into other computers *and people* better than we ever could. Just like cultural evolution, digital evolution is the process in which a form of technology helps us to proliferate our genes. This doesn't mean that we are literally proliferating our genes when we use computers, but the *drive* to use computers is the same drive which pushes us to mate, to paint, and to get ahead in our careers.

Evolution is Really About Information

If we take a good look at evolution, we see that it is really all about information. Genes are bits of instructions stored in DNA. Natural selection grades the effectiveness of the organism based on their DNA. If the instructions are good, the animal can pass on its genes. If not, the animal doesn't and that set of instructions eventually dies out too. This happens all the time — not just when animals go extinct, but when they evolve. The former version dies out and a new version replaces it. That's what evolution does — it replaces old information with better information. Biological evolution works on the information encoded in DNA — it reshuffles it and selects some genes over others. Cultural evolution works on information in the form of thoughts, ideas, and beliefs, which are proxies for the information in your DNA.

The technology of biological evolution for humans is sexual repro duction. We take the information in our bodies and mix it with the information in someone else's body and get a new version. The technology in cultural evolution is also about taking information and mixing it up by interpreting and rethinking it, and then trying to spread it as far and wide as possible. Digital evolution represents the next leap. It is a technology that does the same thing as cultural evolution using the interpreting, rethinking, and spreading power of computers. In all cases, natural selection is the judge as to whether the new version is an improvement or a dud. You still with me? At first we had information in genes, then technology led to information outside our brains, and now technology is bringing information into computers in order to distribute it faster than is humanly possible.

It's as if these bits of information live a life of their own in a digital world interacting with the rest of the world as if they were our physical selves. My comment on your Facebook post is now living among the comments of others. In a weird sense, I am *there* with all those other people having a conversation about your vacation pictures. "Looks like you all had a great time. Sooo jealous."

"We lived on farms, then we lived in cities, and now we're going to live on the internet!"
— Sean Parker in *The Social Network*

But really, we don't live on the internet — only our memes do. We still live in a physical world and success is still measured in physical units. Digital success must translate to physical success. The Great Loophole will still let you feel a sense of pleasure when you do well digitally, but you will come closer to the real reward when you see results in the physical world. That means digital systems are only useful to the extent that they improve your resources. We all want to be a star on the internet, but really we want that stardom to translate to a better car, better mate, cooler social club, and bigger house.

Improving Resources

We are at the very beginning of our digital evolution, which, over time, will allow us to improve our resources in ways we can hardly imagine. I would love to get by on less sleep. I would love to have more energy in my daily life. I would love to be more efficient. I would love to not have to do all kinds of menial work. I would love to know if there is some disease brooding in my body. I would love to never be a victim of crime or violence. I would love to never get lost while driving. Speaking of driving, I would love to never be in another traffic jam. And I would love for all my friends and family to enjoy these wonderful things too. It's all on the way, and digital evolution will be the technology to take us there. Why? Because we want these things. "I thought you said our *genes* want them." Touché. Our genes want us to wake up feeling alert and ready to kick ass. *I* want to sleep in until 9 o'clock and don't care how long it takes me to wake up. But my genes won't have it. There is work to be done and I need to be rested and ready to go.[46]

Yes, you will become part robot. But hey, we got used to flying at 30,000 feet too, didn't we? Flying stinks but the destinations are awesome. Maybe you don't want to have your whole life digitized, but, again, nature doesn't care what you think you want. Your genes will want it if it helps them spread, and the rewards your genes will bring you will far outweigh the costs. I'm not saying you should hook yourself up to a computer, I'm saying you're *going* to. At least be happy that we get a say in it — that stuff has to look nice and feel comfortable. One thing should be clear to you by now, and that is you can't stop progress. **Progress is determined by our genes, by Mother Nature herself, not by us.** If she wants you to feel energetic and stay healthy and never have to turn off your car in the middle of the freeway, then she'll make you feel good about it somehow.

---

[46] I know, "Sleep in till 9 o'clock? That's not sleeping in!" Just wait till you have kids, man. If it's 9 o'clock and no one's woken you up yet, then something horrible has happened.

Before you get too worried about the coming change, let's grab the resources list and look at the benefits, many of which are already under way. Here are the resources about to be improved by computers:

Physical appeal – Digital systems will not change the way you look, but they will change the way you look online, essentially changing how much of the world sees you. How many people see your picture but don't actually see you? If it's a lot, you will enjoy the benefits of a digital existence. Of course we only put nice pictures of ourselves online! We'd be crazy not to. You think I let people put any old picture of me on Facebook? Hell, no. My own profile picture is from 1997 when I had a full head of black hair and rippling muscles[47]. . . . And people don't always use an image of themselves, but rather a scene or an image of someone else. We can make these virtual representations of ourselves as appealing as we want, and that attractiveness carries over to our actual person.

Location – They call it the "death of distance" – people no longer need to be within physical proximity to work together or play together. You can hire and be hired by people all over the world. This means your location matters less than ever. The internet is leveling the playing field for those who live in the middle of nowhere. In a sense, we can live in the coolest places in the world and even many at the same time, but those places are digital places.

Assets – Can computers help you make money via an online business? Yes, but that's not what I mean. Setting up a business in this booming industry is no different from Henry Ford setting up a car manufacturing company. But it is different in one sense: Virtually every obstacle to success has been lowered by the digital age — the costs, the distribution, the people to do it, etc. People from anywhere in the world can become rich by starting companies online for a fraction of what it used to cost. Now, not to burst your bubble, but the competition has gone up to meet the opportunity. It's still not easy, just easier.

---

47 Yes, I'm joking . . . it was 1998.

But digital systems actually do make money in a very real sense. They allow us to move money from one place to another with a click of a few buttons. You can take money from one place and move it to a better place where it will grow more — be it a bank account, fund, or investment. How important is that ability? Hugely important. It doesn't mean all those with a computer are getting rich; they are just better off than those without a computer. Money that can't move is worthless! A big feature of money is that it transfers better than the cows you sold to get it. If you can't transfer it to someone else, you might as well smoke it. Digital systems have mega-expanded the possibilities of moving money and have thereby increased the value of that money ever so slightly.

Energy and physical ability - At the moment, the most powerful digital health machines are the ones in hospitals keeping people alive and checking for diseases. They take input from your body, digitize it, and report on it. They can check for cancer; they can create full color 3-D representations of your innards. Some will even act on your body — like respirators. Many of these machines cost millions. But they won't for long. Apple might have something cooking right now that will produce similar feats in the form of a wrist watch.

Technology companies are hot to hook up systems to your body — to read information from it, digitize it, and report on it. They will process the information, come to new conclusions about your health, and eventually act on your body to improve its functioning. What are we talking about here?

I think we're all kind of sick of doctors, no? You come in only when you have a complaint, you get a half-hearted solution, they send you back out into the streets, no better than you were before but with a significantly lighter wallet. It's not their fault — they don't have the time to spend with you and if they did, you wouldn't have the money to pay them. Some rich folks have dedicated doctors, but the rest of us don't. But, as usual, evolution is bringing the goods to the masses. What we can look forward to is a dedicated doctor — one who has full, continuous access to the inner workings of our bodies and does nothing but spend all day analyzing that information and finding ways to make our bodies work better. Computer systems will be able to check your hormone levels, blood sugar levels, and all other kinds of things; recommend improvements and eventually execute those improvements. There will be little robots you can put in your blood stream who will report back on your health. Not tomorrow, not next year, but it's coming. Can you imagine that we used to have to go to the doctor, give a sample of blood, and wait one week for the results? "Did you take a horse to get there, Dad?" How old-fashioned!

If we're lucky, we've had a few moments when we've physically felt like a million bucks. The rest has been about a buck-twenty. Digital evolution will pave the way for many, many people to feel much, much better than they ever imagined, more often than they could imagine. You'll still have to exercise and eat right, but all the little wrinkles in your physical experience will be ironed out. I'm very much looking forward to that, myself.

Social Network – Today, as always, your social network makes up a big chunk of your resources. The difference today is that there are more people you can be in touch with. Despite all the criticisms of Facebook, LinkedIn, and Twitter your social network really is improved by these social sites. It's all about knowing people and getting support from those people. You can be in contact with 10 real people and get a bit of support from them, or you can digitally be in contact with 1000 and get a very tiny bit of support from each. And though some of those digital friends do absolutely nothing for you, through the Great Loophole, you still have a drive to get their attention and are still rewarded for doing so.

All those people who spend hours a day on these sites are doing nothing less than bolstering their social network resource in an attempt to improve their GPP. Never thought of that, did you? You thought they were just exhibitionists who get off on sharing their every thought with the world. . . . No, they are trying to proliferate their genes as nature instructs. With each tweet and/or post they hope to improve their social standing in the eyes of the ever growing number of people on the other end of an iPhone. Those eyes may never have actually seen the tweeter/poster, but in their mind an image of that person exists and now their social network has improved ever so slightly. That's all the difference natural selection needs.

The digital future of our social lives involves us knowing the whereabouts of friends at any moment in time. Knowing where they are now, where they have been, and where they will be. "Going to Mexico? Peter can tell you about some good hotels, he was there last spring. Oh wait, he already blogged about it, here are the hotels he mentions in his blog. These are the double room prices for the same hotel for the dates you were there. 'Stay away from the chicken fajitas,' says Peter." We're almost there. We just have to connect the dots a bit more.

Knowledge – The benefits are obvious. . . . OK, I'll go into it really quickly. Digital systems essentially have taken the road paved by cultural evolution and turned it into a runway. When I was a kid in the '80s we had to look things up in encyclopedias. Encyclopedias are amazing! This incredible breadth of knowledge spread out across 50 volumes of hardback books. Consider the fact that all that knowledge went through the minds of human beings, and then they took it, summarized it and mass printed it, and some guy came to your house to sell it. Are you, like me, in awe of what an achievement this is? If not, can you take a moment to think about all the thousands of years we've spent learning and how a group of people did their best to condense it all into a product that fits on a bookshelf and is indexed for quick access to the parts you need to see at any specific moment. . . . Astounding. Then came the internet and it shat all over this great achievement and then used Volume 1 to wipe its backside . . . and then took a picture of the gruesome deed and emailed it to its friends, via the internet. . . .

The internet is so much better than encyclopedias, better than anything we could ever have imagined. Anything we want to know is like "click-type-click-click" and we're there. The internet and all the knowledge on it is the ultimate evolutionary weapon. If a group of 50 people had the internet and the rest of us did not, we would so be their slaves right now. Every baby born would be *their* baby. Thankfully this is not the case. Mass competition and a few key policies see to it that this deadly weapon is spread out among billions of people.

Digital systems are knowledge enhancers. They produce knowledge for us *when* we need it. Some stuff we can hold in our heads - like "Duck sauce is delicious". But other information pertains only to certain moments, and we only need to know it in those moments. A knowledge enhancer might be a navigation system or a spell-checker. I don't have to know how to get to Freiburg now or how to spell "guillotine," but computers can help me out *when* I need to know these things. And these things do contribute to GPP. Ever been late for a job interview because you couldn't find the place? You're 0 and 1 before you even get through the door (that's bad). Slight differences mean a lot to evolution. If you spelled "acheivement" wrong on your resume, like I just did now, you may not have even gotten the job interview. Slight difference yet again. These are important to evolution, so tools that give you the edge add to your GPP and help make you a successful human being.

Consider today's navigation systems. Getting from A to B has always been a part of human life. It used to be us just us hanging out in a group of people on the African savannah, so there weren't many places to go since there were only so many miles you could cover in a day. When technical advancements allowed us to travel further in a day, they gave us both the possibility to do more and also the possibility to get lost more.

We used to have maps — big pieces of folded paper. And you'd be like, "Honey, should I get off here or the next exit?" And your wife would awkwardly unfold this huge sheet of paper, and try to find where you were at that moment. And you'd be like, "Hurry, the exit is coming up any second." And she'd be all, "You could have said this two minutes ago." And even if you wanted to get off at the exit, you can't see because the map is blocking your right-side mirror. Well, now we provide our location via zeros and ones to satellites in orbit which take that information to show us where we are on a digital map in real time, and then take our desired destination and tell us which roads we have to take to get there. Some will even replace heavily congested roads with roads with no congestion. You can even get Snoop Dogg or John Cleese to tell you where to go. Soon enough you will be sitting back reading the paper or masturbating to pornography while your car drives itself.

"That does sound intriguing, but there must be downsides." Is it intrusive? Have forests been destroyed? "No" on both counts. We love digital things. Some of us shudder to think of life without these systems. We have not adopted any technology that is too much for us to handle, since despite the fact that our genes are in control, we are the ones who pay for this technology. The people who make the technology still have to sell it to human beings — not some unit of replication that doesn't consider ease of use or how it looks to the rest of the world. Genes get what they want and so do we. The people with navigation systems will ultimately have an advantage over the people described in the map scenario, and thus, digital evolution will see to it that they go extinct.

Computers are also helping us in the monumental task of finding a mate. Some people are the type that go out to bars, easily walk up to complete strangers, initiate a conversation, and ultimately land a new mate. The rest of us are waiting for lightening to strike in our social encounters. But dating sites seem to be leveling the playing field for the mate-less introverts among us. You digitize yourself by filling out your profile — which amounts to transferring your entire physical and mental being into the confines of 10 multiple choice fields and 10 free form fields. Based on this now organized, semantic, digital information, computers can start looking for a mate for you. Because finding mates today is not only hard, it's inefficient. You have to meet someone in a social setting and talk to someone in enough depth to get a sense of whether a) they like you, b) you like them, c) if they are even interested in finding a mate. You might meet a pretty girl who just broke up with her boyfriend, and she seems to like you and can hold a decent conversation, but then you find out a week into dating her that *P.S. I Love You* is her favorite movie . . . and you're back to square one.

But the same cultural reasons to reject someone can provide reasons to consider someone, too. There are dating sites limited to people who hold certain religious or political beliefs. And most sites (I'm assuming here) understand that a person's tastes and hobbies are a good measure of compatibility, hence they ask you to list your "favorite movie," "favorite music," and "favorite activities" to let potential mates know up front whether you two are gonna be fighting over which radio station to listen to in the car.

Since having a family can be considered half of the fulfillment pie, finding a mate is a big deal, and those who have typically had a hard time finding a mate benefit tremendously from dating sites and even newer mobile versions.

In this example, we can see the overlap of knowledge and intelligence. We want to know about potential mates, but really we just want to know where to find them and make a decision about them. Computers are helping with both.

Intelligence – Part of information retrieval is knowing what to know when, but also knowing *why* we want to know something. Computers have gotten good at "thinking" what bits of information enrich the bits you are actually looking for and helping us make decisions based on that information. Cumulative knowledge is thereby more easily attained. If you look up what the weather will be tomorrow, it might tell you "chilly and rainy." This simple insight leads to a whole series of other realizations: you might need a jacket, the trip you plan to take might take longer because of traffic. Every winter I forget to leave 15 minutes earlier since I'll need to scrape my windshield and deal with slippery roads. Digital systems start to put two and two together and help you to make decisions. "Wear a jacket tomorrow and leave 20 minutes earlier for work." It's not curing cancer, but it *is* intelligence.

Remember that intelligence is tied to a specific goal. Digital systems will consider your goal of being an effective human being and provide you with relevant information to help you make decisions and even suggest a few decisions for you. Amazon's website is a fine example of this. Yes, they are trying to sell you stuff, but consider for a second that they are simply trying to help you achieve your goal. Your goal might be to learn about coral reefs. So the system will look at what your peers bought, it will take your age and other purchases into consideration. Even though you are looking for books, it might show you movies that you hadn't thought to look into. It'll even show you deals to help you save money. It is all trying to improve their revenue, but they only succeed when they help you attain your goal. And it is all fully automated. Someone programmed a computer to think like this with a bunch of "if-then-else" statements and now you have this little digital helper there to assist you in attaining your goals. In this sense, Amazon has actually made you more intelligent. You now have a leg-up on people who are walking through physical book stores searching for books on coral reefs. And a leg-up is what you and your genes want.

Thoughts - One of people's favorite activities is thinking — not in terms of goals and solutions, but about more subjective matters. "What does it all mean?" "What do other people think?" "What kind of person am I?" Some psychologists might say that bad answers to such questions are the root of much suffering in the world. Good, capable, otherwise healthy people think bad thoughts about themselves or about the world and act accordingly. There is nothing wrong with their brains, but their thoughts stink.

Mental health — or the lack thereof — is a big problem. As we move further away from the Sweet Spot, our brains are being stretched to sometimes intolerable degrees. Our thoughts can sometimes be vessels of destruction, turning otherwise fine human beings into confused lumps of flesh who flap and flail at the slightest sign of trouble and break more things than they fix in the course of their lifetime. But the problem is really "all in their head," and the solution is right around the corner. The only real limitation so far has been the lack of professional help - because all these people can be helped, they are just not getting the help they need. Just like doctors, psychologists don't have the time to dedicate to one person as such a person would require. Pills have become the commercial solution to this problem. A psychiatrist can spend 30 minutes with a patient, prescribe some pills and the pills work 24 hours a day to fix the problem. But prescribing pills, at the moment, is like trying to put out a fire with a big gust of wind. It might work, but water would be better and less crap would get destroyed in the process.

Not to belittle psychologists, because they are a good bunch doing important things, but a lot of what therapists do is just sit there and listen. They guide you, they stimulate you, good ones call you on your bullshit, but *you* do the actual reconsidering of your thoughts. The research seems to show that psychotherapy from a distance is just as effective as the face-to-face method. Some people even prefer it. What's more, digital psychologists — systems built to emulate the questions and empathetic response of real psychologists — are starting to gain ground. What does this mean for you? Well let me ask you, how great would it be if every thought you had was the best one for the moment, if every emotional response was the most empowering?

One of the other downsides of psychotherapy — and I can tell you this from experience — is that it's not always there when you need it. Feeling like crap *now*? Having a fight with your partner *now*? Well, you will have to wait six more days until your scheduled appointment to actually deal with it. What we need, and what the digital age will provide, is not only on-demand counseling and coaching, but also alert systems which will be triggered when destructive patterns start to appear. How will they do this? I don't know. But I can tell you that bad thoughts can be recognized by machines, and it is just a matter of time before we have these machines installed to monitor our thoughts and corral them into a better place like a cowboy bringing in a stray calf. "A machine monitoring my thoughts!? Not a chance." Of course not a chance, but maybe you'll try it and like it. If not, then you won't have to use it. But if you consider how so much bad stuff in the world is caused by people who don't have a good grip on their thoughts, you'll quickly see how much better the world could be if everyone in it was having a really good day.

The future holds some crazy stuff, which can be scary, but remember, people would also have been terrified in 1910 if you told them about how our world works today. In fact, we are much better off now. Technology is not a bad word, it's the *best* word. And digital evolution is the next great batch of technology that will help us get everything we want, and then some. We need the planet to be able to support eleven billion people, we want more trees and nature preserves and more time to spend in them, we want more money, more love, better feelings and emotions, less violence, less discrimination, less pollution, less suffering, more free time, and more exercise. Quanology reasons that all these things can only come from more technology, not less. Quanology also shows us that technology is not just something digital. Of course, there would be more natural areas in the world if all the people were eradicated, but we should all understand that this is not an option. As long as there are even 100 humans on the planet, the only course is technology. Human society is *all or nothing*.

Jump Right In

This whole chapter was supposed to be like two pages. . . . What was the point of all this? Not to sell you on digital evolution, because it will sell itself. The idea is to show you that even with all the bells and whistles of our connected, wired society, evolution is still the only game in town. And like digital evolution, and all the *Evolutionary Brothers* before it, our genes are driving us to get an edge over others through technology. This is not something we can stop, nor should it be something we want to stop. It will all fall into place in due course. Like any other change, merging with digital systems will be uncomfortable for most in the beginning, but evolution will see to it that eventually everyone gets on board – just like they all got on board with writing. Imagine how upset a monkey would be if you gave him a pen and made him write a letter to his zookeeper when he wanted something. It would stink for him in the beginning, but if he got the hang of it, it would open up all kinds of doors for him. "Would you add some strawberries to the daily fruit basket, please?"

Our goal, as always, is fulfillment. Genes want to be replicated, but the Great Loophole has allowed us some wiggle room to do our own thing. We are like the CEO who, despite needing to improve profits every year, can still use the company jet for personal reasons. To repeat the mantra, we will be fulfilled when we do those things that traditionally led to higher GPR in the Sweet Spot. Improving our GPP (our personal value) is a great way to feel good, and digital evolution opens up a world of opportunities to do just that. You saw the quote at the beginning of this chapter: Gary Vaynerchuk wants everyone in the world to go to his funeral. If his funeral is digital, then things just might work out for him.

What a mosaic we have painted living in these three levels of evolution. Using the Four Recipes in these three layers, we have created an incredible world that you have to stand back and look at every now and then. I hope you've enjoyed the tour. Each layer of evolution offers a new world of opportunity to do well in life. To be fulfilled, our bodies and minds will have to occupy all three — the biological, cultural, and digital worlds. What does that mean to your everyday life? Biologically, you should raise children and spend lots of time in high energy places with high energy people. Culturally, you should innovate, learn, share your ideas, and express yourself. And digitally, you should get connected and start to augment yourself with digital technologies. Let them help you to be a better version of yourself. Because that's what this is all about, remember? You.

5 | Personal Development

About You

Understanding evolution — the very force of life itself — can help you be a better you. It's been a long ride, but at last, we're about to put the "U" in evolution.

We've been through a lot, you and me. Murder, sex, love, betrayal, drugs, and piles of cash. This *tour de vie* has shown you how the world works and how natural selection is still the controlling force in our lives. It was often long and complicated, but it was the truth. The better we understand the truth, the better we can move forward. To quote John Lennon: How can I go forward, if I don't know which way I'm facing? We want to move forward, but in order to do that we first needed to understand where we are now, and we needed to know what *forward* really means. And now, it's your time. The rest of the book will be all about *you*.

Or should I say all about your Gene-ish Self: your family, friends, colleagues, and even the people on the other side of the world, who are working to help improve your life and those of your family, friends, and colleagues. We're all working for our genes, but we're also all working together. On average, all the positive sum interactions we engage in all day every day result in an increase in our GPR. Nature is after high GPR. Nature rewards us if we get it. And we get it by helping ourselves and those around us. Ultimately it is you as an individual who needs to act and work in this direction — to point your boat downstream on the Great River. The better you do this, the more fulfilled your life will be.

So, go ahead and kick back for a while. How's life? But really think about it. How's your life? Think about past events, current events and where you want it all to go in the future. Did you recognize yourself in the past 200 pages? Did some things trigger memories or thoughts of the future? What about the inner experience? What emotions have you been experiencing on a regular basis? What about your daily activities? Do you have some variety or are you stuck in a routine? How do you feel through it all? Do you feel you are in control of your life, or do you feel like you are on a treadmill and can't get off?

As I've lived and worked alongside all different kinds of people in my life, I've noticed that they are all usually full of hot air in most circumstances. When it comes to questions like these, people are not completely honest with you. They're not lying, but they're not being open either. That's part of the game, as we've seen — keeping up appearances. Those who are honest may look weak, and that's bad for business.

I've come to understand people better and better over the course of my life, and at some point something occurred to me: I don't know anyone who doesn't have real problems. And don't *you* pretend. I've been in therapy and studied myself and others for a long time now. You can't fool me. I know what the score is. No one I know is free of some emotional pain, or has their life 100% in order. No one. And you may think, "Dude, you've got to get to know some better people," but that's not the answer. They are good people. It doesn't matter how successful or fulfilled you are, no one makes it through life unscathed. After reading this book, you should not be surprised by this. Getting hurt is part of the game. But realizing this will help soften the blow and allow us to focus less on the problems and more on the goals that inevitably produce them.

If we look back on our lives, most of us have some regrets — we wish we had done some things better. Maybe our past has led to a present we're not too crazy about. So, you think about your relationships. You think about your job. You think about the commitments you are in and dream of getting out of them. If you're lucky, or should I say *successful*, you also have things you are really happy about. Maybe you love your job, your friends, and your family. Maybe you feel in control and are able to get more of the things you like, and eliminate those you don't. Either way, every now and again, it's a good idea to have an honest look at your life and see if you're on track. In my experience, most of us are not, and pretty much everyone could be doing better in some way. Who among us can say that everything is going perfectly? . . . Exactly. So we develop ourselves.

This practice is *personal development*. Call it self-help, if you will. It should provide a set of goals and then help you on your way to achieving them. You need to be a capable person and then you need a clear and actionable plan. Because there's this one thing we all want. . . .

The Quan

That's right — that mythical thing called the Quan. A life lived fully. A life that gets more than it gives. A life full of love, admiration, and glory. Great sex, good food, and luxury. Sounds like a fantasy, and it is, for the most part. In this book, we have brought it down to earth — down to biology. Quanology comes in to give clarity, direction, and action to getting the best life possible.

Quanology ultimately provides a new success paradigm. It's not about money, it's not about achievement — **it's about doing what nature wants from us**. We now know what our purpose is on this planet, and we can go about fulfilling it. In doing so, we will enjoy a greater sense of meaning. For me, that meaning was not at all obvious.

Progress means developing. It means creating things that other people use, supporting a family, and enriching the lives of others. These things feel good because nature wants them done, and not because of some noble plan, but because these things dissipate energy well. So, we do them, and nature writes our paycheck by making us feel fulfilled. Underneath it all is a bunch of selfish genes — nothing but nucleic acids — but the end result is something beautiful, just as millions of colored dots thoughtfully placed on a canvas can make a beautiful painting.

You may just be a small part of that painting, but it is your part and it means the world to you. So, you try hard. As we look around, we see that some people are doing well in this world and some people are doing really poorly. Those who are doing well are respected for their contribution to society. They have a set of friends who are also respected. Their relationships with friends and colleagues are positive and cooperative. They have enough money for all the things they want to do. They are attractive and have a good sex life. They may have a prosperous family. And on the inside, things are going well too. They like themselves and are confident in their abilities. They learn new things all the time. The emotions they experience are overwhelmingly positive, and they have a deep-rooted belief that the world is a good place and this is all being done for some higher purpose.

How cool would it be to be one of those people? I'm working on it. I'll assume you are too. So, let's take a quick look at you and how you can be a better version of yourself.

Who are you?

You are a gene replication machine. Through the course of hundreds of millions of years, you have developed into quite a good-looking human being. You're smart. You're funny. You dress well. You make people smile. They make you smile. Some things make you happy, some things upset you. You like some people, you don't like others. You want people to like you. You can be horny. You can be driven. You can be lazy. Sometimes you don't know what you feel but you want it to stop. You can feel lonely. You can feel connected with other people and, when the moment hits you, you can feel connected with the world.

Natural selection is behind all of this. It made you look good so that others would mate with you. It made you smart to figure out how to navigate the world and all the other people living in it. It gave you emotions to teach you and steer you toward the right behaviors: the behaviors that would lead to better gene proliferation. It made you personable and self-aware so as to be more effective around others. It taught you to pick a team and dislike other teams so you would be stronger in your quest. It made you horny so you would have sex. It made you driven so you would win. It made you lazy to make you reconsider how you are doing things. It made you feel funky so that you would change your ineffective behaviors. It made you connect so that your bonds would be stronger with others. It made you think that you were close to God so that you would never give up.

That is you. That is me. That is all of us, forever.

Yet we are all different and fairly unique. We have the same goal and a certain number of strategies, but we all start off with different circumstances which weigh heavily on our lives. We think of "who we are" as a mental function, but as far as nature is concerned, we are more than that. We've seen that "who we are" is the sum of our resources and technologies. Without the technologies, we are grunting cavemen. Without the resources, we are no longer people but some other form of life.

But now, let's focus on that mental aspect of you. Your mind is most directly related to you as an individual. Your body is still very important, but it has undeniably taken a back seat to the brain — especially when compared to other animals. The mind is the biggest contributor to who you are and how you will experience your life.

What is all too obvious is that some brains function better in the world than others. We care about our brains because we want them to help us get a fulfilling life. We want them working optimally to do this. We call those people with optimally working brains "mentally healthy." How mentally healthy are you? To what extent do you feel your brain is helping you and not hurting you? I'll quote Green Day again: "Are you your own worst enemy?" Or does your brain rise to the mental challenges around it?

What do mentally healthy people look like? Here are some key traits:

- They feel in control of their lives.
- They are in control of themselves.
- They experience positive relationships.
- They feel as though they have plenty of time, money, and energy to give to those around them.
- They know who they are and like who they are — there is no effort to "be someone else."
- They are more out to give than to get, but the world is giving them plenty.
- They feel their lives have a sense of purpose which motivates them beyond any personal obstacles.
- They feel no need to put down or manipulate other people to get ahead, because they themselves are on cloud nine, and no one can kick them off.

These people are few and far between. You'll know them when you see them like a red horse at a cocktail party. If you feel doubt about someone, then they are probably not the real deal. The majority of us are, in fact, filled with quirks and less-than-perfect character traits. In our attempts to increase our GPP, or at least not diminish it, we do not always act like angels. That's OK, we're not angels — we're humans. Animals. . . .

As for the rest of us, we've all got a little something. Look through the DSM-5, the official list of mental illnesses, and you can be sure that you have a few disorders on a low level. But a low level disorder is not enough to take you out of the game, and with a little practice and perspective, you'll be just fine.

What's clear today is that our brain is being stretched. Changes to our physical environment are met by changes to our body through biological evolution. For example, the world demands that we move out of the trees and onto the plains, so we develop a slimmer pelvis and start to walk upright. But the changes in our culture happen too fast for the matter in our brains to change with it. So, we get the Great Loophole. We get addicted to chocolate and get scared watching horror movies. These things are bugs in our software. You can report them to tech support, but they probably won't get around to implementing any fixes until the next release . . . and there is no planned next release.

As such, you are stuck in an awkward position. A stranger in a strange land. It would all make much more sense if we were grunting and living outdoors in small groups, but then there would only be a few thousand of us in the world and the genes would not be pleased with that result. They are happy now with 7 billion of us squirming and fidgeting in a modern world with nuclear weapons, vaccines, and double lattes.

But, if we're honest and objective, you're doing OK, too. Life will always be full of struggle and dissatisfaction, but also love and joy. And you will always be trying to feel better through all of it. Hopefully, this book has taken away a degree of confusion for you, as it has for me, and helped provide a railing for those moments you're not sure where the ground ends and a pitfall begins. Chasing the wrong things are genuine pitfalls and misunderstanding the world will cause its fair share of unneeded misery. It's an increasingly confusing world only because we are covering our real purpose with mounds and mounds of misleading thoughts. Our real purpose is still there at the foundation, and we still get the best life by fulfilling it.

Your Purpose

If you have some big plans for your life, good. Now evaluate them against the backdrop of natural selection. Are those plans supporting your genes in the world? If not, you will go unrewarded. If they are supporting your genes, or even if your genes just think they are being supported, and you find success, you will live a fulfilled life. I would argue that is the best any of us can hope for. The rest is riddled with nonsense and leads to intellectual dead-ends.

Our purpose on the planet has been articulated many times, but not enough from the standpoint of the living. We hear things like "we are here to be happy" or "we are here to love." I never felt these explanations were good enough and developing Quanology has shown me that, in fact, they are not good enough. Other explanations such as "we want to grow" or "we want to contribute" are more accurate but still need a foundation. I hope I've given them just that.

If all this talk about genes is getting your feathers ruffled, I understand. Seeing fulfillment as the reward for success can seem like a bit of a drag. I mean, we don't want to be content, we want to be happy, right? Look at commercials of this or that and all the people are constantly smiling. "Can't a brother have a laugh in this bitch?" Your concern is noted, so let's see why happiness and fun are not getting the traditional props here. Really, it's a matter of design.

You are simply not built to be happy for any sustainable period of time. Sustaining a high point, any high point, for a long period of time is just not natural. If you were constantly happy, you would not want to change. And nature feasts on change. Happiness is great, but it's not a motivator. Companies don't want to be happy, nor do countries, nor do sports teams. They want to be fulfilled. If you are confused about why people do what they do, it's because we are all subconsciously seeking fulfillment, not happiness, but we consciously *think* we are seeking happiness, and that messes everything up.

We all want to trade in bad emotions for fun emotions, and to a degree, we can. But the bad emotions are there for a reason, and as long as they exist in other people, they have to exist in us in order to navigate the world effectively. Can you imagine being the one guy or girl who never gets mad? You would never make it in the real world among humans who *do* get mad. You wouldn't fit. And if you really want to be honest, we *are* experiencing more positive emotions today than in the past. Sure, there is a lot of tension out there and uncertainty, but there has always been *something*.

Maybe it used to be sickness or poverty or war, but now it's job security, health benefits, and working in a job you hate. People are experiencing better emotions with every generation. Those who will tell you that the past was much better are deluded. They themselves are afraid of life and subconsciously try to coax the rest of us back to a state which exists only in their heads. When we hate the present, how fondly we remember the past.

In Steven Pinker's *How the Mind Works*, he mentions two reasons why happiness is a lost cause for humans. Firstly, we always compare ourselves to other people. In the information age, we get to see other people better than ever. Remember *Lifestyles of the Rich and Famous*? If not, you know *MTV Cribs*. I personally hate watching those shows since, despite being curious, I feel down about my life afterwards. *Why don't I have 3 Bentleys?* 'Cause I damn well should! Objectively, humans have more reason to be happy than ever, but when we compare ourselves to others in better positions — social media makes it easier than ever to do this — we shoot that objective reality down. Somebody somewhere will always have it better than you in some way. And doesn't that just burn?

The second detriment to our happiness is the fact that we, as good subjects of evolution, adapt to our environment. Whatever situation we find ourselves in becomes normal. It's great if you're able to feel happy being poor and oppressed, but on the opposite end of the spectrum, extreme wealth can also start to feel normal. The extra happiness boost you get from any gift that comes into your life — even winning the lottery — will eventually wear off.

So, why bother trying to be happy? Because *happy* is fun. It feels great. It looks great. It works well with others. You should strive to be happy in your life, just like you should strive to be wealthy and physically fit. Happiness, like wealth and health, are resources. They are not the end goal, but they will help you get your end goal.

Though people are experiencing more happiness in the world, it's still not the end game. Happiness is a fleeting mental state, not something profound. You can train yourself to be happier by changing the things you focus on, the questions you ask yourself, and the way you answer them (which we'll talk about more in a bit). The results can come quickly. Positivity follows much the same path. Of course, you need to be more positive because it will get you better results in the world. So will dressing well and eating healthy. But positivity is not your goal, just like looking good is not your goal — it's simply a technology you implement to improve your mental resources.

You should be happy because happiness is a resource. So go hang out with friends, and play sports and Zen-out on a beach in Thailand. But, remember, you're here to achieve a goal. Those yogis who want to just sit around and meditate all day are freaks, and I'm not afraid to say it. Fine if they want to do it, but the rest of us need to work. If we all did nothing, slowly but surely we would devolve back to the Sweet Spot, and you can expect all the hardships and strife that come with doing whatever you have to do just to survive.

Thankfully, it doesn't look like that will happen. Most of us are working and are still *motivated* by the drive to pass on our genes. That may still be confusing to you — you, who gets up and goes to work every day, never giving so much as a thought to biology. . . . But through this new pair of glasses we see that the most important, most fulfilling, most beautiful parts of our life are rooted in just that: our biology. Biology is, after all, about life.

Goals in Evolutionary Terms: Kids and Contribution

We've learned new terms in this book — GPR, GPS, and GPP. And now that you know where they come from and why I had to define them, let's put them in layman's terms and look at how they show up in your daily life.

GPP = your self-worth
GPS = personality and behavior
GPR = success

You want to be successful and now we know what success is, but how do we translate it to a human life without counting our genes? How do we create a life that is fun, fulfilling, meaningful, and *human*? In a nutshell, it boils down to giving. **You are successful to the degree that you give.** Giving means progressing the goal of nature: creating things and bettering situations and people. It means giving life to children and/or contributing to the lives of others through technology. You can do one or the other or both. They are both hard, but fulfilling in their own way. And despite sounding "Disney," they are both completely gene-ish.

Life in the Sweet Spot demanded that we pass on our genes to raise our GPR. Those with the highest GPR got rewarded the most be nature. Not only did they have the most high quality children and the highest standing in their tribe, they *felt* good. They were the most fulfilled and had the highest GPP. Why? Because back in the Sweet Spot your genetic proliferation potential was more closely related to your genetic proliferation rate. When you had sex, you were more likely to have babies than today. When you did something good, everyone knew about it and gave you your props. And you knew *everyone*.

The Great Loophole has changed this, and now those with high GPP do not necessarily have high GPR. We use contraception when we have sex but we still feel the natural rewards of sex. We adopt children but still feel the natural rewards of raising them. We study astronomy and still feel the natural rewards of contributing technology to our group. We share information with complete strangers but still feel the natural rewards of passing on knowledge to our kin. We work out and we still feel the natural rewards from a body that thinks we are working hard to fight and hunt. Unlike drugs and overeating, these members of the Great Loophole have no downsides (unless all that bonking gets you into trouble). These are just the things we need to do in our lives to feel fulfilled. GPP and GPR are most closely linked with respect to raising children, developing social relations, and contributing to society, and I'll summarize them with the term *giving*.

"When you say the goal of life is to give, aren't you copping out and falling into the same touchy-feely trap that many self-help books fall into?" No, Pam! It's there. Of course, it's more complicated than just giving, and the intention is as mundane as the laws of motion, but this is the shortest, most effective way to put it. You need simplicity to focus and there it is. Go murder and rape and see how that feels. Dedicate your life to money and power and see how well your goals of eternal life through genetic proliferation are fulfilled. **You won't do too badly, but you can do better**. I mean, you can be a bad person if you want to, but you won't enjoy it, and neither will the people around you. There may have been a time, long ago, when things might have worked out for you, but no longer. So all you preachers and fear mongers can relax. We don't need laws to tell us what's right and wrong, we have each other, and we have our own internal compass.

Those with the highest GPR and the highest level of fulfillment are givers. If we look at celebrities and famous titans of industry, what underlies their success is giving. Even that ham, Larry Ellison, is a giver; he gave us a fast relational database. When I first saw him on a cover of a magazine acting like some slick billionaire a-hole, the word "giver" would not have been on my list of adjectives to describe him. Quanology has changed that — he is a giver. So are Bill Gates and Warren Buffet. They may be rich, but they got that way by giving. Bill Gates has given us so much stuff. Whatever you techies have to say about Windows, you can't deny that it changed the world. He gave us computing and he is more fulfilled for that contribution. Now, he's done with that and wants to keep getting rewarded for contributions, so he's got a foundation that's doing all kinds of good work in the world. Again, through giving, he is fulfilled.

But you don't have to be rich or famous to be fulfilled, you can also be a good parent. Good parents are mega-givers. Good friends, too, but perhaps to a lesser extent. If you're none of those, there is something missing in your life.

In life, there is really one big question to answer. Before I had kids, I thought it was "*to get married* or *stay single*." **But I later realized that**

the real question is "*kids* or *no kids*." Everyone wants to be in a serious relationship at some point. If they are not it's because something is going wrong. Either they consciously deny they want a serious relationship or they just can't get one or hold on to one for very long. But not everybody wants to have *kids*. Nor should they. You can do a lot without kids.

What about career? If you're not having kids, then career is not a question — you need to have some sort of plan for cultural contribution, which typically comes in the form of "structured work." In lieu of a family, cultural contribution can also take the form of supporting friends. I have seen people in this situation doing quite well for themselves. They have a close knit group of friends and they are actively doing things together. It's organized and it's intimate. But since you'll still need to make money, those friends won't be enough — you'll need a career.

So, your life boils down to a couple of options: to have kids and/or to have a good career. Let's look first at kids.

Kids are hard work — more than I ever imagined. They overshadow everything else in your life. Despite that, there are many people who spend their whole lives wanting nothing more than to raise a bunch of kids and have a big, happy family. They may be on to something. If it didn't result in such a messy house, I might be one of those people.

Most sane people think babies are cute. Nature has designed us to feel this way and, for that matter, it also designed babies to be cute. When you see people with babies walking in the park on a sunny day, or you visit them at their house, it all seems cool. You don't see them losing many hours of sleep, or freaking out about how small their house now seems, or how unfit their neighborhood is for child rearing. In their "keeping up GPPearances," they only tell you how wonderful it is to have kids. And the next thing you know, you're romanticizing about having kids yourself. Hell, what else are you gonna do?

In a very concrete way, having kids gives you something to do. Most of us are pressed for time after we leave school and join the work force, but then you wake up one morning and realize that's all there is. "I go to work and come home and that's it." You might find little hobbies, hang out a bit with friends when you can find the energy, but your life is pretty focused on work. You want something for yourself. The answer may be to have children.

Now, much research suggests that people with children are less happy than those without children, but I would like to add some Quanological perspective to that conclusion. Children are an investment and you can get a good return or a bad return, just like any investment. If you are less happy with your children, it may be because you are screwing it all up. Kids will be most rewarding when you do well with them. In the moments when I think, "This is all a mess," I see it's because I'm not being a good father.

People with children and people without children are also like two different species. Not much changes in life when you get married, especially if you were already in a serious relationship before getting hitched, which you probably were. But having kids changes your life in every way, and therefore, people with kids live in a completely different world from people without kids.

Yes, you have little time as it is, but in a sense, having children frees you from many other responsibilities. If you're a lady person, maybe you will start working less. If you're a man, maybe your wife will start working less. . . . Once you are pregnant, the world opens up to you like it never has before. Kids are good for society and so society is there to help. Family members usually spring to your aid, your employer cuts you some slack, and the government protects you. Random people on the street suddenly display some extra sympathy, remembering when they were up to their necks in diapers.

With that new child comes fulfillment. Your house will never be the same again. A good night's sleep becomes a thing of the past and you may breakdown every once in a while and question why you had kids in the first place. But they are wonderful, aren't they? Absolute magic. They are so damn cute and they are developing all the time, constantly giving you new things to ogle at. And your parents are happier with you now that you've given them grandchildren. Life can feel very complete with children. You'll have a whole new perspective about what's important and what's not. I can honestly say the best moments of my life were nice moments with my kids that I barely remember any more. I only feel the warmth I felt, but I don't remember the situation. I can only remember feeling *big*, feeling whole, feeling connected. I feel this new sense of responsibility and importance in the world as I take my family for a drive or to the store or on vacation. The only other thing that comes close to that feeling is profound accomplishment. And that's what kids are: an accomplishment. But you don't have to be a star athlete or prized physicist to have kids. For most of us, it's the most accessible way to feel fulfilled and accomplished. In a very real sense, you are accomplished. You've passed on your genes and have done what nature wanted you to do. She is most grateful.

As I alluded to, there is one big catch . . . you have to be a *good* parent. Being a good parent not only means having a more loving relationship with your child, it means creating a child with a high GPP – that's nature talking fulfillment. We *do* expect results from our kids (kids, your parents *do* expect results from you). That's only natural. So, it's all well and good if you and your kids are close, but it's worth more and feels better when they succeed in life. There are a million ways to mess up a kid and only a few to help them feel secure, confident and driven. If you chose to be a parent, be ready to treat it as any other skill in life. It'll be hard work, but the payoff will be great.

No Kids Allowed

And then there is a life without children. . . . Before you think I'm suggesting that everyone go out and start a family, let me explain that having children has become less important to your life than it was just 50 years ago. We are firmly living in a time where our cultural contribution means more than our biological contribution. You can have a great life without kids that's meaningful, enjoyable, easier on the wallet, with more sleep *and* more sex.

As we've seen, cultural contribution means higher status which adds up to a better quality of life for you, and better prospects for your genes in the form of better mates and better resources for your family. "Sounds great, where do I sign?" Hold on. At the core of cultural contribution is giving. You invent things, you build stuff, govern a town, organize a music festival, start a business, and ultimately what you are doing is giving to others. It's all gene-ishly motivated, but the customers of your business will be better off for your goods, the attendees of the music festival will have had a great experience, your invention will help humanity, etc., etc. And you will be praised by those involved, you will be paid, you will be better known as someone who does good things, and you'll feel great about it. Accomplishment feels great. Giving feels great. Just as with the kids, you have done what nature wanted and she gives you a heartfelt pat on the back.

You want to have a fulfilling life and that means following one or both of these two paths. Maybe you've already gone down one or both, maybe you haven't done either and it's time to pick one. You'll want to consider what feels right for you, and which path best matches the personality you have. Some people really don't want kids and decide to focus their time on culture and friends. On the other end, there are many anonymous men and women who are focused completely on their family. They are anonymous because they haven't really done anything that we would know them for. The maternity ward at the hospital knows them and the schools know them, and that's all they need.

Like I said, even if you have kids, you'll need to be involved in society, make money and earn some degree of respect and recognition if you want to get the good stuff in life. If you don't have kids, no problem, you can focus entirely on that contribution.

Choosing a Career

Career deserves its own special section. Kids are like, "That's not fair—why does Career get a special section and we don't?" 'Cause I said so, that's why! But if you must know . . .

Remember that nature doesn't care if you survive, only if you pass on your genes. But you can't pass on your genes if you don't survive and you won't survive without satisfying some basic needs. We need food, water, shelter, and air. Two of those cost money. Three if you don't live near a clean river or water fountain. A long time ago, our job was to hunt and gather. Now, we make money and buy what we need. But this section is not here to talk about career as a way to make money. It's about the same thing as the last section: doing what nature wants.

If you have a career, you make money. But you only make money when someone feels that your contribution is more valuable than the money he is spending on it. Crazy exceptions and criminality aside, there is no making money without contribution. Someone believes that you are improving their GPR and they pay you for it — in the form of money, status, recognition, or whatever improves your GPR. If you want your GPR improved, you'll have to improve their GPR first.

Think of a career as nothing other than a vehicle to contribute culturally. You want to do good things for people and be recognized for it. Your career is a cultural vehicle for your genes the same way a marriage is a biological vehicle for your genes. Biologically, you are limited to a few quality children. Not so in your career. **Your career can be scalable, and that's why it gets its own section.**

As anyone who's ever started a business knows, a key part of what makes it interesting is scalability. Investors will ask, "Is it scalable"? or, in other words, "Is there a limit to how big it can grow?" If you are the only person who can do the work, then the answer is "yes" — you can only work so many hours in a day, so that becomes the limiting factor in how big your business can become. If you can hire people to do the work, then the answer moves toward "no," there is no limit, because you can hire as many people as your cash flow allows. That's a simplification, but that's about how it goes. A web-based company, like Facebook, has the potential to reach a global customer base and deliver unlimited service. What's more, while the number of users and revenue increase linearly, the costs diminish. That's scalability!

You'll want to eventually have a career that is scalable. If you can only work 40 hours a week and only make money when you work, then your scalability is crap. If you can delegate work here and there, automate tasks with technology, then you're on your way to scalability.

Ultimately you are a brand — that's where the cultural contribution gets tied to you. Virgin is Richard Branson's brand, and in becoming so culturally successful through name recognition and by giving to so many people, he has improved his GPP. All businesses do this on some level. Not all are so good at giving or at linking the business with the person behind it, but they try. Johnson Plumbing does it. And so does Mercedes. And yes, you, as a humble employee at a company can do it, and get a higher level of GPP than the unrecognized employees who don't really contribute much. They are the anonymous henchmen that die in James Bond and Austin Powers movies.

Not everyone is made to be a business mogul, but **no matter what you do**, you can increase your value by taking pride in your work, being prepared to learn and innovate, and enjoying the rewards you create for your customers. This is an entrepreneurial approach to your career. The more in contact you are with your customers, the more you will feel that natural fulfillment. The more you innovate and deliver the goods, the more rewarded you will be — not just financially but by nature itself.

Becoming scalable also means eventually moving into the employer role, where you are the ring leader who delegates the work to other people. Those other people can be systems as well — maybe you sell a service and that service is some automated digital system or even a product. Markus Frind owns a dating site that makes millions but he only has a few employees. Instagram only had nine employees before they sold to Facebook for $1 billion, so you don't need to run a huge company to be successful. What you do need is to work hard.

I never wanted to work hard, but things happen. You get out there and you see how hard everyone is working and realize there is no way you will make it without working hard. Working hard by definition sucks. Otherwise it would be called "working easy." But this is where nature makes the selection. Who is willing to work harder? It's not all about hard work, of course — there is creativity, focus, and confidence and all kinds of other factors. But working hard is the price of entry. Even executives who we think sit around and do nothing all day are actually always managing their plans, goals, and social positions, so don't let exteriors fool you. It will be through success and the meaning you attach to it that you will get rewarded by nature for that hard work.

Yes, you'll need to work hard, be creative and competitive, but at what? Your work should fit you. Some guys are made to score touchdowns and some guys are made to sack quarterbacks. Some women are made to be mothers, and others are made more for the workplace and still others the object of desire. All of these flavors can be successful. Success for you starts with understanding who you are, so feel free to invest in self-discovery.

This is your life. Your family and your career are the vehicles to success. You can be as creative as you want. If we let the terms of GPR, GPS, and GPP go, all you need to do is be the best version of yourself you can be, all the while investing in your family and friends and/or your work. Excellence in both will provide you deep fulfillment. However, by definition, excellence is not easy or quick. It takes focus and effort.

Quanology is about our ultimate goals in life; it's not about how to get money. But, there are a few goals that show up again and again in the self-help books that sell like hot cakes: money, sex, health, happiness, and love. While they sound like good goals, we often don't understand *why* they are good goals, and thereby, fail to reap their full rewards. As heavenly as it sounds, you cannot be in love your whole life. You cannot be happy your whole life. Love is the tool, not the goal. You can be healthy and beautiful your whole life, but so what? You can be rich your whole life, but if you don't understand what wealth is for, it's like having a private jet that you never fly in.

We are animals, and just like all the other animals, we have a purpose. That purpose has not faded with designer jeans and luxury apartment buildings; its execution has simply become more intricate and therefore we don't see it as clearly anymore. I've spent the last few hours of your life showing that, underneath it all, that purpose is still there. Hopefully, in breaking down our human strategies in terms of kids and careers, you can see that family and cultural contribution are the respective deliverables on the two layers of evolution.

Past generations have measured men by how much wealth they had, how many people they ruled, how many wars they won, or how many women they slept with. Women were measured by their physical attractiveness. **Today people are measured by what they give to others**. It's not New Age, it's not religious, it's natural. In my humble opinion, it's beautiful.

Getting to Work: Evolution in Your life

We've talked about our goals and we've talked about how a successful life means investing in those goals. You want to do well in life. But investing can have a good or a bad outcome. There is always an outcome — life is not just an *experience* or a *journey* or *what you make of it.*

Personal development, or self-help, as a subject, is there to help you score well. But for better or worse (and for me it was worse), personal development doesn't have a clear end goal or provide a map of the world. You cannot determine success if you do not know the goal, and that is what Quanology provides. Only after I understood what a *person* is and which direction *development* points toward, did I really feel comfortable with the process of personal development. Now that I have that information, I can jump into life fully motivated. I hope this book has shown you how the world works and which goals are worthy of achieving. And now you want to achieve those goals, don't you? That means becoming the best person you can be.

The best self-help books offer a program, not a bunch of "truths." *Quanology* is more of a "how life works" book, not a "how to kick ass at life" book. Of course, when you know how life works, you should expect to be better at it, but life is a skill and skills are best developed through practice, not reading. Maybe I'll throw a program into the next book. . . . For now, we're just going to talk about your life and look at what Quanology means to you as a person.

Wanting a Better Life

First, we have to agree that you want a better life. It is common among average people to pretend that everything is great in their lives and that being open about needing to improve is a sign of weakness. But everybody could be doing better at something. It's just you and me here. . . . OK, it's just you—I'm not really there, so you can feel free to be open about your life while reading this next bit and think about what you really want as if you could really have it and be honest about your failures. I know that even this is a stretch for some people. Even when they are all alone, they will not admit to weakness or lack of perfection. It could open Pandora's Box and screw up their whole routine, screw up the lie they sell themselves every day that things are going OK. But every now and again, you need to do it if you want to get anywhere. Without pain and suffering there is no progress. Too bad, right? But that's how it goes. Most people hum through life in mediocre mode, never feeling any real pain, but never feeling that great either. The only way to get out of mediocre mode is to feel pain, and taking a good hard look at your life is just the shot in the crotch to do it. Get ready!

Motivation often only comes through feeling enough pain to justify making a change. We evolve biologically only when those who do not adapt *die*. Motivation is crucial to improving yourself. There is a big switch in your brain that needs to be flipped. If you're not sure whether it's been flipped, then it hasn't been. Some people have had it flipped as children as part of their upbringing. But every successful person in the world has flipped their motivation switch. Though other people can try to motivate you, only you can flip that switch. That will happen when you feel a stark contrast between what your life could be like and what it is now, and you not only start to believe that you deserve better, but that you can get it and you'll like it.

If it helps, and it should, you're not just doing it for yourself, you're doing it for the whole world today and the whole world tomorrow. The better you build yourself up, the more of that technology you pass on to others. Your family and friends will experience the benefits most and the people on the other side of the world, maybe only slightly, if at all, but it's very real. When you die, and you will die, those technologies, those self-improvements, will live on in the people you have touched in your life. And that number of people is bigger than you may think.

Try to allow yourself to make mistakes along the way. To want to be perfect from the start is unnatural. And what have we said about that? So, get yourself a program, since this isn't one. See what feels right and start getting to work.

And work you will. . . . Someone once said you can work hard and enjoy the fruits of your labor or you can skip the work and suffer the consequences. Both are hard, but the first one feels good. In your life, you should strive to be a high value person. This sounds obvious but there are enough people who strive to avoid work. I used to be one of them. I thought being successful meant getting by with the smallest amount of effort on my part. But I had to change, and it was really my mind-set that had to change first. Life is hard and some people say "screw it" and try to escape. They still need money to survive so they become cheaters and freeloaders. But that's hard too, and it feels like crap. The fact is there is no escape aside from death. And you don't want to kill yourself, there's too much cool stuff to see and do in life.

For me, accepting that life was hard work was not a given. I saw all the people around me working hard, and I didn't want to be like any of them. Not because they were jerks or miserable, but because I didn't want to do what they were doing. It wasn't me. The greatest place you can arrive at is knowing your calling in life — that big thing that you were born to do. If you haven't found that, you can at least start on the path that Quanology paves for you by becoming a better you today. You can start by getting to know yourself. Are you going to be a parent? Or would you rather focus on your contribution to the world? If you're not sure about this or even more detailed questions, you can still start developing yourself as a person — improving your resources through technology. Get that make-over. Get in shape. Start developing your social network, move into a better environment. So that when it becomes clear what you want to do, you'll have the potential to get it.

Despite the conflicting advice you'll find in the self-help book shelves, you undeniably need to improve yourself as a person to do well in life. We agree on that, we just argue the best way to go about improving yourself. As this book has been all about nature, I want to let nature give her take on things. The Great River is flowing in society and in your life too. Let's look at six evolutionary principles and what they mean to your life.

The Principles of Evolution in Your Life and How to Use Them

1. Fit Into Your Environment

The Principle

We start with the most basic principle of evolution: fitting into an environment. Natural selection is not about "best," it's about "best fit." All the plants and animals currently on the planet are the best fit for their environment. Take them out of their natural environment and they will have a hard time. They will have to change, and if they don't, they will die. You are driven to get high GPR, but you will only succeed if you fit into your environment. **Your natural, unstoppable tendency is to fit into your current environment.**

As we've seen, this world has steadily become a nicer place in which to live. We are working together more and fighting less. This creates an environment where pleasant people fare better than unpleasant people.

Those who seek to help others are winning out over those who are only out for themselves. Nature is seeing to it that our dark sides are going out of style. People are not consciously choosing to be nicer, they are simply fitting in to their environment. If we want even less bad things, we need to continue to create environments that stifle them.

You too are fitting into your environment, wherever you are. Describe your environment now. What kind of people are around you? What do they value? The fun starts when you see all the things around you as characteristics of an environment and know that your behavior is being subconsciously steered by that environment. Then you are better able to take control and change that environment to something more to your liking. And when you do that, your behavior will naturally follow. **We can call these environmental pressures "expectations."** For the most part, we meet our expectations. Go watch *Trading Places* if you're not sure what I mean.

Take a guy from a bad neighborhood with bad expectations and he will most likely meet them. Put him in a high expectation environment and he will probably do the same. Let's look at Mike Tyson. He grew up in a horrible neighborhood in Brooklyn and turned to crime and violence at a young age. Why? Because he is a bad seed? If you listen to the guy, he's super sweet. In his 2007 documentary, he's very open about the fear he felt when strangers would attack him and try to rob him. His environment expected him to be tough and mean and to take whatever opportunities were presented, even if it meant stealing. So, he did. His environment was filled with thieves and thugs and he adapted to become the same.

He was discovered at age 14 at the juvenile prison where he was serving time, and was abruptly taken in by professional boxing coaches who had a different set of expectations. They expected him to work hard and be disciplined. During the week he lived upstate with a middle class family who treated him well, so he became that sweet guy around them. On the weekends, he went back to Brownsville and immediately went back into his thug life. Ultimately, he spent more time in the environment of training and middle class living. Within a few years he won the Gold Medal in boxing at the Olympics and, obviously, went on to become the youngest, and arguably greatest Heavy Weight Champion.

Both environments may have been unpleasant. Both environments have their own set of problems. It must suck having to fend for yourself on the street with drug dealers and crime bosses, but it probably sucks to wake up at 5 AM and work out full throttle all day, too. The point is, you adjust to everything. New situations become normal in a fairly short amount of time. In that sense, you don't have to be afraid of the hard work that lies ahead. You can go from one extreme to the other but still feel you are living on the same planet and inhabiting the same body. But the expectations of your environment determine what you do. You need the outside pressure to create those expectations. Not only is it too *hard* to stay in an environment and act differently than it expects, it's not *natural*. You were built to adapt.

The Application

> ***"God, grant me the serenity to accept the things I cannot change, the courage to change the things that I can, and the wisdom to know the difference"***
> – Serenity Prayer, Reinhold Niebuhr

> ***"You know what my father told me when I was growing up? 'Look like you've been there before'. . . . Even though you don't have it, you have to dress as though you got it."***
> – JB Smoove

The best chance you have for success is to choose the environment you want to be in, and find a way to get there so that the expectations of that environment will help you become that better version of yourself. This is the real value of getting into a good school and making the "right" group of friends. That environment creates high expectations that you will naturally live up to.

Of course, many people do rise above their environment. Some, despite coming from the gutter, make something great of themselves. Others living in the lap of luxury fall hard to the ground. But they are the exceptions and not the rule. You might be inspired by success stories of those who rose above their desperate circumstances to make something grand of themselves and think "I can too" but if you want to play the odds, bet on being a slave to your environment.

If you take a look around, most people are meeting the expectations of those around them. Not all, some do much more, some do much less, but the rest of us are doing what the people around us tell us to do. If you come from a high GPP environment, then that environment will expect you to also be high GPP. You may resent mommy and daddy pushing you to study so hard, but you will be pushed to have a high quality of life, and to work hard and excel at what you do. If you don't come from such an environment, but you want to excel in life, then you may need to take the initiative to find another environment that will have high expectations for you.

We have the control to be and do whatever we want, but our environment, even our mental environment, heavily influences what we do on a daily basis. **If we want to change things in our life, we will, therefore, need to consciously change our environment**. This goes against much of what the self-help industry tells us. Even psychology is starting to try to get us to change our own behavior through sheer will-power. And we can to a certain extent, and we should certainly try. But if our environment remains the same, statistically, little will change. I personally know very few people who have simply said, "I'm going to be different now" and have dramatically changed their lives. What I do see is everyone around me meeting the expectations of their environment. They do this without thinking. I am a lazy bastard who hates to work out. I need someone watching over me to force me to work out. Luckily I have this.

I have a friend who I jog with regularly, and the teacher at my Ving Tsun school gives me subtle guilt trips if I don't show up to train.

Now, it's not easy to change your environment, but being conscious of it helps. So, let's take a look at lining up your environment with the things you want in your life. And remember what those things are. In your environment, you want things like:

- Positive people.
- People who want or have the same goals as you — whether those goals have to do with family or cultural contribution.
- People who have a similar GPS to you — are they entrepreneurs, artists, thinkers, accountants, lawyers, athletes?
- High energy, large communities like cities; or lower energy, smaller communities like rural areas.

People, places, and things

What elements of our environment can we change? **The most important element of your environment is other people**. They determine so much of your behavior. If you are around people who contribute to a bad environment, you will need to distance yourself from them. Maybe not cut off all contact tomorrow, but start to mentally tag people as a good influence or a bad influence. Maybe you've already done this but never realized the subconscious impact people can have on you. If their behavior starkly contrasts with the environment you want to live in, then that's no good for you.

I'm not just talking about negative people, I'm talking about people who are very different from you. When you are around them you will naturally try to fit in, and this is also no good. It will take you away from who you are. If you're the artistic type and you hang around accountants with totally different values and opinions, then you will feel out of place and subconsciously start to take on their values. There are tons of activities, events, and places that attract people with specific interests. You would do yourself good to find those that interest you and are in line with your goals and personality. High quality people from other cultures and walks of life are not to be avoided, for they can be a benefit as well. Just keep in check that natural tendency to be more like whoever you are around, if you don't want to be more like them.

Where you live also has a huge influence on your life. Cities have more energy and higher expectations than rural areas. People move to cities to get more out of life. But wherever you live, you want to make sure you feel satisfied by the things and people around you. I live in a Dutch suburb that I'm not too crazy about. It was a mistake to move here, but it's not so bad and now the kids are settled and like it. We have a nice house and wonderful neighbors, so we're not moving. In that sense, it's acceptable. But I'd rather be in Amsterdam. Now that's a cool city with cool people and interesting activities. It's an hour away by train and 45 minutes by car, so what I can do, and what we do do is go there every now and again and hang out. We borrow that energy. We put ourselves in that environment and allow it to shape us.

Within our suburban life we have found people to hang out with who feel the same way we do and who like the same things. We all cheer each other on and keep the expectations high. Look around at the people, places, and things in your life. Which ones fit and which don't? Take some time to look for upgrades.

Leaving people

When things aren't working, you may need to leave some people behind. Now, it's one thing to start doing new activities or move to a new city to get a new environment, but changing the people in our lives is a different story. This is hard. We love the people around us, and upgrading our lives while the ones we love maintain theirs can amount to treason. There is this thing called passive aggressive punishment where you subconsciously give someone a guilt trip for stepping out of line. It happens in everyone's life regardless of GPP. Maybe your wife is starting to go out more, and while you consciously tell her you are happy she is "getting her social life back," subconsciously you hate it and all of a sudden become annoyed with her lack of attention to the housework. It's totally natural, but it's a barrier to overcome. There are more than a few belief systems which urge you to break from your current friends and even your family. I'm certainly not saying that. Your family loves you, but they may very well be holding you back. And while you can't just leave them in the dust, with the technology of today, you can join up with a different peer group on the side. Think of it as having "a peer-group affair." You are married to one peer group but you are seeing another one on the side.

What are you doing that is so out of line with your current peer group? Maybe you are trying to be successful and the people around you are slackers. Maybe you are trying to start a serious relationship and your friends are all about quick hook-ups. Maybe you want to get in shape but you hang with over-eaters and heavy drinkers. Maybe you are surrounded by stock brokers and you want to do landscaping. **If you are not meeting their expectations anymore, there will be conflicts.** They will naturally push you to stay in line while you naturally try to please them. It's time to find some new groups of people to hang out with. Easier said than done. Your peer group will expect you to be there with them in the future just as now.

This is one of the greatest obstacles to success. In talking about beliefs we saw that some beliefs are bad for you, like "I'm not that smart." These limiting beliefs inside you say you belong right where you are, but there are also beliefs on the outside, like "John is not that smart." External beliefs can become just as integral as our environment to who we are. But if you want to get significantly more than you have and do significantly more than you do, you are going to have to alter many relationships in the process. It is unavoidable.

Now that we've learned about GPS and GPR, we know that we are all motivated to proliferate our own genes and that, as beautiful as our relationships are, they are there to serve a function and are gene-ish in nature. To break those relationships or change them into something different is therefore not treasonous, it is a part of the deal. You're not a bad person for doing so. Some relationships you maintain out of obligation and some out of real love. The obligatory relationships should never have been there in the first place and the loved ones will only be served by your doing well in life.

Getting new people

Enough about your luggage, let's talk about your new peer group. Do you know who they are? Like any product or service you want to sell, you need to know your market. So, again, you ask yourself the question "Who are the people I want to hang out with?" I've had some good friends in my day, but I've never really been able to say "These people are like me." Maybe I never knew who I was or was just too picky. However, I can see the things I like and the way I think reflected in popular culture. The psychological demographic that I do connect with is not so much a group of people hanging out somewhere, but a group of thoughts floating over communication streams. That is my community and I need to be true to it.

In the process of writing this book, that focus has kept me in line. I'm not writing for my parents or in-laws or even my wife or kids. I'm writing for that group of thoughts, held by people and expressed through popular culture. How is this a technology? A long time ago you couldn't choose what beliefs you held. We are continuously moving to a freer mental environment where you can feel and think the way that best fits you.

We know that family is a wonderful thing and there is much fulfillment to be had in the harmonious relationships of those people who share our genes. That is not to exclude others, it is simply to confirm our suspicion that blood *is* thicker than water. A dysfunctional family like the Bundys from *Married with Children* can be a pain in the ass, but those asses share something very precious. There is a bond there that transcends the daily squabbles of a family forced together uncomfortably. If that communication goes smoothly, even better.

But we also know that there is much fulfillment to be found in the sharing of a cultural experience, and the ideas and thoughts therein. This is where you can create your cultural family and spread a cultural identity that is strong enough to persevere into the future.

Your mental environment

Lastly, your own thoughts are a crucial element in your environment. Yes, your environment will guide your thoughts, but there will be moments when you consciously try to control your thoughts independently from your environment. What are the thoughts running through your head on a daily basis? Are they creating a good or bad environment?

I'm saving a more detailed explanation of this for later in the book. For now, understand that there is an environment in your head. Bad thoughts will not do well in a positive mind, and good thoughts will not do well in a negative mind. But your mental environment can be changed, too.

2. Change Gradually

"No plan survives contact with the enemy."
— Tom Holden from West Point Academy

"Everybody has a plan until they get punched in the face."
— Mike Tyson

The Principle

Change through natural selection occurs imperceptibly from one generation to the next. Small changes are enough to create small advantages and over time they result in new species. Plants and animals only make small changes because they are not sure what nature wants. Nature doesn't even know what she wants, she just knows if the change is a good one or a bad one. If you, as a plant or animal, make too big of a change, you run the risk of changing too much for the worse and killing your chances at surviving. See, the thing is, you can change in more ways for the worse than you can for the better, so it's best to hedge your bets and change *small.* If it's not too bad, you won't die. If it's for the better, you can try to keep going in the same direction in the next generation. This is how plants and animals make their way through the process of evolution: They follow the laws of physics and the theories of math, and they get to new places through small changes.

The Application

The progress of your life, if done right, is an evolution. Preferably, things should progress at a pace that is just slightly uncomfortable for you. If things are too slow, you won't accomplish as much as you could, and if they go too fast, you won't be ready for them and the weight may prove to be too much to maintain. Look at someone like Brittany Spears or Eminem. Things moved too quickly for them, and they weren't able to handle it. We all think it would be great if our lives went from boring to incredible overnight. But in reality this would be a curse.

> ***"My father always used to say: 'What doesn't kill you, makes you stronger'. Until the accident."***
> – Jimmy Carr

The same applies to failures. Some self-help gurus tell you to "fail big," but this is of course nonsense. You want to avoid mistakes, but we know they are unavoidable. At least by making small, calculated moves, you can limit the damage. A big failure is one that has a big impact on GPP and could take a long time to recover from. If you are rich or very young, then some mistakes are not a big deal. You have to take risks, but start small and learn bit by bit. The more you learn, the less risky those next bigger moves will be. In the physical world, you need to start small.

The mental world can be a different story. Big shifts in your mind can occur more quickly and that's cool. When I discovered how the world works as I've shown you in this book, my mind changed forever. When I realized I could control my moods, I was never the same.

When you know where you want to go, you can start right away with small steps. Knowing the final destination will motivate you to get going, but the journey needs to be made one baby step at a time. The average day of your life should be imperceptibly better than the day before. Yes, there will also be "aha" moments when all of a sudden a bunch of tiny changes amass into something more noticeable. These are wonderful confirmations of progress! But nine days out of ten, you may think you are getting nowhere. As a year goes by, maybe you also don't really know what has changed since you don't remember where you were a year before. Keeping a journal is a great way to track your progress and look back on the person you were the year before. You're trying to reach a place where things you once thought were scary or out of your league are now normal. This happens step by step.

In his book, *The Lean Startup*, Eric Reiss describes how startups need to work as quickly as possible to produce a Minimum Viable Product, a minimal version of a product that can be shown to the world in order to get feedback on its validity.

His rationale goes like this. Want to create a killer website? Start with a minimalist version and go from there. Don't spend months working on the perfect version and then show it to the world. The world is too complicated and it's changing too fast to know exactly what will work when you fire your gun at this moving target. So, go slow. Get feedback from nature. In the case of a tech startup, nature=users. This is evolution. MVPs are what nature needs for natural selection to work. There has to be something to judge and that something always comes in the form of a small difference. Big differences are too risky for the simple reason that they take lots of investment to create. If you can create bit by bit with little investments, you can change direction and never really lose too much time or energy when you screw up . . . and you will screw up.

In daily life, I use the term Minimum Viable Action, or *MVA*, to mean an evolutionary step in the direction of your goals. You start looking into new education opportunities, you look for a new pair of jogging shoes, and you search for networking events in your area. These little actions change you and they amount to slight adaptations that natural selection can judge. What you want to do is set aside time every day to take an MVA (sounds like a pill, right?). What MVAs do is allow for slow, gradual change. Since they don't take much effort you don't run the risk of investing too much energy in something that doesn't cut the mustard. It could be that the direction you are heading is not the direction you really want to be moving in. Maybe the product you're creating is total crap and no one wants it. Maybe your approach to finding the perfect man sucks.

What's more, MVAs won't freak your subconscious out with lofty goals. If we close our eyes and imagine a grandiose life in stark contrast to our current life, our systems resist. We are always adapted to our current environment, so if you propose big changes, your body and mind naturally resist. Try telling a fish that he has to live on land next week. It's a step up, but he's so scared he has a long strand of poop hanging out of his butt. Take small steps in that direction, and you'll avoid the fear that comes with any big change and limit the excuses you'll think up to stay where you are.

Scheduling

All projects need a schedule and that means allotting time to work on them. Obviously, it would be best if you could work on your new improved life 24/7. A few lucky people in the world have this luxury. The rest of us have other obligations. The fuller your current schedule, the harder it will be, but it's certainly not impossible.

Before you start to think about how much time you have for your project, please take a minute to appreciate the things you already have. I have a wife and kids and most of my day is dedicated to them. At times, I get annoyed that I have less time to dedicate to other things, but that is when I'm not appreciating the fact that my kids are the most important thing in my life. Sometimes I have to slap myself and remember how good I have it. Gene-ishly, they are part of my plans too, and I need to schedule time each day to help them out with their lives. Piano lessons, homework, sports, English lessons, etc.

Some of us have things in our lives that should not be there — a bad relationship, a bad lifestyle, or bad habit. These things also take time out of our day that we could be spending on our projects. If you can identify one of those things in your life (like watching too much TV or drinking too many beers), then consider adding it to your project as an item for discussion.

There is a technology in this world called "getting things done." There are people who are experts on how to make yourself do things that are important but perhaps painful, or complicated. We each have our own methodology, since I can't imagine that one size fits all. But there are resources out there for you, and whatever your problem, you can bet that someone else has it, too. With the internet, it's never been easier to look into how to deal with problems big and small.

My little system, if you're interested, is as follows. I have short little lists, and I make myself spend five minutes working on each item. Of course, you basically can't get anything done in five minutes, but the point is, I say to myself that it's OK to just spend five minutes, and then I always spend more than five minutes. The hardest part of working out is putting on your sneakers, so for me, spending the five minutes is the equivalent of putting on my sneakers. As Tony Robbins put it, "An object in motion tends to stay in motion and an object at rest tends to stay at rest." The Laws of Motion apply eerily to your life. It takes a lot of effort to get started but less to keep going. Force yourself to get started and be less strict about how much you do, since more often than not, you'll do more than you expected.

3. Invest

"It [investing] is not about maximizing return, it's about maximizing quality of life."
– Tim Ferriss

The Principle

Evolution is a system for investing. **To do well in life, you need to invest**. That's all animals and plants are doing. Their success is measured by the return they get. A chameleon invests in the ability to change color. It spends energy in that direction which could have been spent in another direction — like running faster. But if changing color pays off in the form of higher GPR, then the investment was a success, and soon enough all chameleons will be able to change color. And so it is with humans. We expend energy doing things like having kids, pushing for a promotion at work, trying hard to get that special guy or girl to go out with us, or spending extra time learning a new skill all to get further in life. These are investments. And you'll need to work hard and make sure they pay off.

The Application

From this day forth, think of all your actions as investments. You give a little to get a little more back. When you're tired, when you feel like sitting on the couch and watching an episode of *Cheers* that you've already seen 3 times, you need to realize that if you work instead, the reward will be greater. Yes, *Cheers* is funny, but it is a poor investment of a half hour - especially on the fourth viewing.

All your actions are investments which will have either a good or bad return. Success in life will mean getting a good return on the majority of the investments you make, especially investments that apply to all those resources we talked about: your appearance, location, physical ability, social network, knowledge, assets, etc. We're not just doing these things to improve the resource themselves, but to get actual results in the two paths: family and contribution. Success in those paths will reach beyond your own life, but it is you who will get joy and fulfilment.

This concept should give you renewed drive and a boost of energy to work hard at all the aspects of your life. If you get the right return, you'll have an easier, more fulfilling time later. Some years ago, the wife and I spent some time in couple's therapy. It was a hassle. It was expensive.

It required babysitters and time off from work. But it paid off. I'm not gonna tell you we never have fights, 'cause we do, but for the most part it's a damn love fest over here. I never knew family life could be this good.

So, start investing. Pick one of the resources on the list — maybe the one you feel is your weakest point. Picking too many will overwhelm you, and that would ruin the small changes principle. So, just one for now. And then pick the family or career path to focus on. Now think of little investments you can make in that resource that would help you do well in your chosen path. I just chose the social network resource in my career path. As a new author exiled in The Netherlands, I have a crappy network. What investments can I make to improve my social network related to my chosen career path?

Coincidentally, I just signed up for an online course. It costs me money, and it will cost me time — not just following the course, but also taking the actions the course outlines. It will be an investment, but if it pays off, it will be a very good one. Today there are an infinite number of options for investing. Do a bit of research up front, make small investments, and get going.

4. Focus

The Principle

All animals and plants have found their niche. They do one thing well. You cannot be all things. A cheetah cannot run fast and swim well. If it tries to do both, it will lose out to the cheetahs that have focused on speed. The competition of the modern world makes this principle even more relevant. There are people like Jamie Foxx who have found success doing multiple things, but their initial success came through focus. After all he's done, it's easy to forget that he started his career as a comedian. He didn't start singing, acting, and being a bad-ass until later in his career. How could he have known he'd be singing on Kanye's album? He didn't even know Kanye existed.

I personally have a hard time with this – like I'm weak and worthless if I can't do everything. But the laws of physics apply to me too, and every now and again, I need to pull back the reins and focus on bringing my current investment full circle.

The Application

If this is also an issue for you, you need to find a way to be cool with yourself. Understand that virtually no one is totally happy with who they are and what they are doing at the moment. People always want more. But you'll have to push that drive aside from time to time and focus. Limit your goals, limit the person you need to be, and limit the things you are doing. You can take on new challenges after you've accomplished your initial goal. So, for now, focus and kick ass at the task at hand.

Go ahead and write down what you want for your life. Is it all over the place or is it nice and crisp? **Fulfillment comes when you shine brightly in one area, not when you do pretty well in a bunch of arenas**. Accomplishment is in itself an asset. Doing well at something — anything — is something you can show the world as a testament to your ability. It's as good as money. You can use it as a springboard for your next endeavor even if it's slightly different from your previous one. Steve Jobs moved from computing at Apple to computer animation. He was able to do this because his accomplishments at Apple opened the door for him, arguably more than money could. Had he tried to do computing and animation at the same time, he may never have been able to spend the necessary energy to make either a success.

5. Use Technology

The Principle

We've seen that life has been a technology game since it began. The animals that best adapt to fit their environment get to live. When we see these adaptations as biological technologies, we see a direct link between animal and human technology. The people who most effectively use technology are gaining the upper hand in life. For me, the beauty of

Quanology comes from understanding that technology is everywhere, and you don't have to be a geek to use it, since it's not just about electronic devices you hold in your hand.

Yes, we have technology in the form of physical tools like smartphones and cars, but we also have technology in our thoughts and ideas, as well as technologies that let us work well together. The individuals and groups who have best used these technologies have done better in life.

The Application

We all need to become technicians if we are to do well in life. But it makes for an easy transition when we see that technology is in every little thing we do, like shopping online to save time and money, learning better approaches to parenting, etc. If you approach life like this, you become a technician — you have taken a tool and used or improved it. In doing so, you have improved your own life as well as improved the world. Good job!

The first place to get technical is in your mind.

The mind

The mind is a great machine whose software can be reprogrammed. You can't upgrade your genes but you can upgrade your technology. As technology advances in computers and medicine and transportation, so does the technology of the mind. Every few years bring exciting new research on how to manipulate your brain toward a more positive and effective state. But the best of it is already there — you just need to be conscious and start practicing. It starts with a look at emotions.

Emotions are a big aspect of life like legs are a big aspect of soccer. Emotions are a *biological* technology at the foundation of our behavior. If we didn't feel love for someone, we wouldn't form relationships. If we didn't feel love for our kids, we would let them fend for themselves. If we didn't feel happiness at accomplishment, we wouldn't do anything positive. And if we didn't feel afraid of heights, we would fall off ledges more often than we currently do. But managing emotions is also a technology. As we've created this world where the reality of an event is up for discussion, our emotions may not always be helpful, so we may need to manipulate them into something that is. Before we do that, we should at least understand them.

Aren't emotions fun? Not all of them. We love to feel love but hate to feel jealous. We love to feel happy and hate to feel angry. And as we get older we often try to consciously experience more of the good emotions and less of the bad ones. But if emotions are just natural reactions to a situation, then really what we want is to fix the root of the problem in order to create more positive situations. If you want to be in love more, be around cool, interesting people. If you want to be less angry, avoid negative people. But bad situations are unavoidable, and bad emotions still serve their own unique purpose. Ever feel angry or sad and actually feel *good* about it in some perverse way? Nature created anger and sadness for a reason and we would be stupid to disregard these emotions. The same goes for the rest of our emotions. They are universal, biologically rooted tools. Jealousy is not learned, guilt is not something awful your parents taught you. The emotions you feel through the course of the day are there to guide you.

Since our society is still driven by the same goals as 50,000 years ago, all the emotions from then are still relevant today. If you're feeling guilty, don't push it away. Find out why. You're in a contract and you're messing it up. Is the guilt justified or should you not have the contract in the first place? Check out these other emotions. Think about when you regularly feel them and why.

- Jealousy
- Love
- Hate
- Anger
- Shame
- Joy
- Sadness

You will notice they all have contracts attached to them in some way. Those contracts are social contracts: "I expect this from that person" or "They expect this from me." Our whole social fabric is based on contracts, and we feel good emotions when they improve our GPP and feel bad emotions when they don't.

Since emotions were created for the Sweet Spot, there is a slight disconnect between our daily routines then and now. But are the situations really so different from 50,000 years ago? Public speaking is considered one of the most frightening experiences in the world, but your life is not in danger. So is it silly to be afraid? Getting up in front of a crowd must have been scary 50,000 years ago, too. It's an act of dominance. It's a person mentally copulating with a number of people all at the same time. He can impregnate them all in one go or be rejected. It's a high-stakes game and fear *is* appropriate.

But in public speaking, fear will screw you over. I've watched some speakers who were scared stiff up there, and their presentations were the worse for it. I was one of them. I stepped out of my body and saw myself trembling and thought, "That guy is terrified! He's screwing up the whole speech!" So, what good is an emotion that screws up your ability to perform? Public speaking, and other activities which are scary, are proving grounds. If you can get up there and be confident, then you can win big. So, yes, it's appropriate to be afraid — and anyone who is new to public speaking will be afraid — but overcoming that fear and making a confident presentation is exactly the way to get a boost in your GPP. We like physically impressive members of the opposite sex because their physique is a sign of strength and virility – that he/she can deal with the physical challenges of the world. In this respect, mastering fear is also a sign that you are able to deal with the high-stakes situations of the world.

The same goes for emotions like jealousy and anger. They are mental tests. We don't want to throw them out; we want to master them. I can be mad at someone but I don't have to lose control. Mastering these emotions and coming out on top in the tough situations are what make the high value people who they are.

There can be big differences in how each of us deals with emotions. **The goodness or badness of a situation is not always clear, so you're not always sure which emotion feels right.** You might be in love with someone and feel insecure and start to experience jealous emotions. In your head, you know there is no good reason to be jealous, but there you go, thinking about all kinds of situations that would warrant jealousy. And that's where psychology plays a role. That interpretation layer is unique to you personally. Some people are more jealous by nature, some more angry by nature, and when those unclear situations come up, one person will experience emotions differently than someone else.

The good news is that there should be a lot of room for improving your emotions in your daily life. How good does life feel when something great happens to you? How good is life when you just did something remarkable? The difference between what you feel now and what you felt then is as simple as whatever made you feel that way. But mental technology allows you to feel that way even without that great thing you just did.

This is the Happiness Switch. It's about changing a few thoughts and a few physiological states, i.e., muscle positions, to change the way you feel. Tony Robbins encourages people to change their physiological state — stand up straight, engage a confident posture and the facial expression to go with it. Doing this changes your mental state, and when your mental state is in good form, your life tends to follow. We can consider this a mental technology.

And often the thoughts we think create bad emotions. We interpret the past in a bad light. Basically, this is Cognitive Psychology. Yes, it's sad that you got a divorce, but if you want to be in love again, you may need to reinterpret some things that happened to you, stitch up the negative emotions that are holding you back, and move on. I know it's easy to write that sentence, but hard to do in practice, and I struggle with negative emotions on a daily basis too. I mean, this book? Don't even get me started. If I had a nickel for every time I got depressed thinking "This book sucks" and "I'm an idiot," I would have a massive stack of nickels. Funny how other days I think the opposite. Which one should I believe? This is my point: both are inaccurate, so choose the good interpretation.

Let me reiterate what I've said before: any method for developing yourself needs to be practiced. Self-improvement is not a "read and learn" exercise, it's a "do and experience" exercise. Up until recently, psychology has been mostly about talking, which is good. If you have problems or are not functioning well, the situation can only be assessed by exchanging information with an expert. Imagine having a problem with your computer and you weren't allowed to look at it or touch it to see what was wrong. Issues needs to be talked out. But only up to a certain point, and then you need to start to fix it and continue with the work it was designed to do. That work is your life and you have to get on with it.

Therefore, self-help programs need to be just that: programs. Tony Robbins has a program. A good one, at that. Definitely worth the money. I'm not up for self-help guru worship, but if there's a demi-God among us, he may very well be it[48]. His program focuses on manipulating your psychological state through movement and body states. Recognizing that the difference between doing well on one day and hating life on another day is just a different state of mind, and he focuses on putting your emotions in an optimal state as often as possible. "Emotion comes from motion," he says, and then he puts your body through the wringer to call up the right emotions — just like double-clicking an icon on your desktop. Double-click and you're more confident. Double-click and you're happier. Have the circumstances of your life changed at all? No, but you become happier and more confident just by changing the software in your mind. That counts for a lot.

"Isn't that kind of cheating?" you ask. Yes, it is. "Shouldn't I expect to only feel good when I do good things?" Of course, but who cares? Why wait

---

[48] Worth the effort to look into him: http://www.quanology.org/tony-robbins

until you do something great to feel great? Imagine the Neanderthals calling us cheaters because we developed more advanced tools instead of evolving an ax-shaped hand. If you want an edge, you have to do what works. Cultural technologies work. And feeling great works. When you feel great, you interact with your environment in a more effective way. Remember the biggest attraction of any environment is *other people*, and those people tend to treat you like you feel. Getting a more positive response from the people of the world is not a little advantage, it's a huge one. Make yourself feel good and notice how other people treat you.

As with all technologies, you are currently using a version that will soon be replaced with a better version. Robbins, and many others, have recommended gratitude exercises for years, teaching people to make notes of the things in their lives that are good, that they appreciate. This is also supposed to improve your mental state. Scientists are starting to study these things empirically. It *sounds* logical: "Think about positive things and you become positive." But what scientists are finding is that you're better off not counting your blessings, but *subtracting* them. You will get better results if you think about what your life would be like if you didn't have all the positive things you currently have. Apparently, research shows that this works better. And so, technology is refined.

Again, this is just one theory. If you can see how important it is to improve your emotional state, put it on your list somewhere and dedicate some time to looking into it. If you can be conscious of your emotional states and close your eyes and remember how it feels to be in a good mood and how cool everyone is to you when you're in one, then you should be able to mirror that state pretty quickly. You won't always want to be in a good mood; sometimes you're mad or unhappy and damn well want to stay that way. Fine. And maybe you really need to for a while in order to figure things out or motivate yourself to change. But the point is there are options available when you're done living in an emotional sewer.

Are results guaranteed? Yes! For all the praise our brain gets, it's truly a pile of dung. That's what I've been trying to tell you. If you do 200 push-ups a day for a month, are you guaranteed results? Of course! It's the same with your brain. It's a physical object and can be manipulated just as predictably.

Positive Psychology is an interesting place to start exploring such topics. This is a very practical branch of psychology that also gives people exercises to improve their mental experience. It's properly recognized by the scientific community. But who cares, right? What works, works. Maybe it doesn't work as well as it could just yet, but it is being approached from a more practical and technological angle than other branches of psychology. Have a Google on "positive psychology exercises" to start putting it to work.

http://www.quanology.org/positive-cognitive-psychology-exercises

Beliefs

> ***"You can trick yourself into believing that life's hard when it's really not. Drama is only as deep as you dig the hole for it to be."***
> – J.J. Grey

Like I said, emotions are a big aspect of life like legs are a big aspect of soccer. And if emotions are like legs, then beliefs are your team spirit. How important is team spirit to a soccer team? Immensely important. They are the yin to the emotions' yang. Just as you need to better manage your emotions, you need to upgrade your beliefs. Whatever situations you find yourself in right now, you have a set of beliefs attached to those situations. Most of us are not even conscious of this, which is how the beliefs get to operate unattended. Regardless of whether it is a good situation or a bad situation, you need to make the *beliefs* empowering ones. Let me give you an example.

My business failed. I spent 6 years on it and lots of money. It was supposed to support my family and give me freedom from the paycheck. But it failed. It hurts to think about it. But what's done is done. So, now what? Do I wallow in self-pity and call it quits on new and interesting ventures? No! But how do you continue after something like that? You slap some empowering beliefs on the situation. Beliefs like:

- I learned so much.
- Everyone has a couple failures in business.
- No big deal, everyone is still alive.
- At least I tried, while everyone else was just dreaming of doing something cool.

These beliefs push me through the pain and help me start the next ventures with renewed confidence. These do not come naturally to me, just like muscles don't come naturally to me. I have to work for empowering beliefs. And I assume you will, too.

Let's think of some situations you or your loved ones might be in and what empowering beliefs can be applied:

Situation: You got a divorce.
Empowering belief: "So glad to get out of a bad marriage and into a new life."

Situation: You have a lame job.
Empowering beliefs: "This is where I can learn how to take pride in what I do." Or "This is just the motivating element I need to kick off my new career."

Situation: You have no idea what you want to do with your life.
Empowering beliefs: "I get to take some time to explore what I really want." Or "Now is the time to build up the vessel that will carry me when it all becomes clear."

Situation: You had a rough childhood.

Empowering beliefs: "I made it through, and now I'm so tough." Or "I have so much compassion for people as a result of what I've been through." Or "Someone else has been through worse and my past is no big deal. It won't determine my future."

Situation: You have way too much to do, and you are exhausted.
Empowering belief: "I have endless energy and everything I do seems to work out well, so it's all time well spent."

Situation: You are afraid you are not good enough to get the things you want in life.
Empowering beliefs: "I'm a good person here to do good things and the world supports me and my goals." Or "Everyone starts at the bottom and works their way up. I'm going step by step."

Situation: You broke up with your boyfriend/girlfriend.
Empowering beliefs: "Now I get to be free and have fun, and spend some time learning about myself." Or "The next time you do something, it always works out better than the last time."

You need to believe these on a deep, deep level, and that probably won't happen overnight. If you're thinking to yourself "I can't do it," just trust me that you can and start off small. Like anything new, it will feel uncomfortable and possibly stupid in the beginning, but you will get the hang of it. If you keep hammering them into your brain and using your body to open the "belief pores," you'll get them in there slowly but surely. Again, it's something that requires practice. I still have moments when I think I suck, or that I'm ugly, or stupid, or socially retarded. In truth, I may be all of those things. But I don't care about the truth when it comes to these kinds of beliefs, I care about what works, so I flip them. And in doing so, I make myself suck less, look better, and act cooler.

Last but not least, let us look at the mother of all beliefs. The belief that has helped armies triumph over foes and has helped athletes win Super Bowls: the belief that God is on their side. If you take it a step further, it is the belief that everything fits, and all that is good and bad in the world exists for a higher purpose, and that higher purpose is not something random or pointless, but something good. Since everything fits, you fit too. Your sorry ass with all your imperfections is an integral part of this wondrous world.

Just because you don't believe in God doesn't mean you can't install this little gem of a belief on your operating system. Yes, you downloaded it illegally, but it's there now, so you might as well use it.

That *feeling* that you are on a mission from God and that the world is tilted in your favor can have a powerful effect on your actions and the way in which you experience life. Business coach and personal development author Brian Tracy calls this feeling that of a "reverse conspiracy" — you act as if the whole world is conspiring in your favor. Every moment of your life, even the failures, have all been leading you to this triumphant moment just around the corner.

Believing God is on your side can help you make it through tough moments. It can give you wings when you feel the weight of 1000 pounds on your shoulders. No, God is not on your side. Even if God existed, which he doesn't, he wouldn't give a damn about you. But since he doesn't exist, you're free to project your own wishes onto his non-existent consciousness. There is no need to dress up for this or to join some church or cult or whatever. You don't even have to call it God. It's something bigger than you. Imagine: Something that rules the world — be it the laws of nature or Aquaman — likes you and is helping you to do what you want to do in this world. Your goals are aligned and you are a team. Take a moment to try that shirt on. Sit up now, fill your lungs, and imagine it — some all-powerful being helping you. How does it feel? If your answer is not "incredible," then you're not doing it right. Keep trying. It'll be worth it.

Just as we evolved to live in groups and to walk upright, we evolved to believe in God. Science has since found this belief to be daft. But it doesn't have to be counter-productive. If you're an atheist and repulsed by this idea, no problem. I am too sometimes, so just let it go. Just be aware that you may be missing a killer app on your smartphone. Food for thought.

Will adopting these positive beliefs make them true? No, but they put you in the right frame of mind. I don't have endless energy, but I accomplish more now than I ever did before by focusing on the positive idea that I can deal with all the things I have to do during the day. Our beliefs pull us down or lift us up accordingly. I used to think that I needed more rest, but that was just an excuse not to get things done. Depression is just code for "scared." Find that thread in your brain that says "I can do it" and feed it like a new-born baby.

If you don't have faith in yourself, your actions will ultimately fail — you won't find a good partner, you won't start a successful business, and you won't keep off the weight. It's the basic starting point for all goals. If you're lacking what you want, then consider the lack of belief as the cause. And then consider a new set of beliefs that don't have to feel true now, but that you can grow into. **This is your spiritual center**. "The world is a good place. I am a good person. The world will benefit from me accomplishing my goals and therefore the world will help me get them." After reading this book, you should have some solid reasons to believe these statements.

This spiritual center will be the extra fuel tank to push you forward. Because life is hard, it's complicated and requires real determination and drive, which come naturally with the right beliefs. You don't have to feel them deep in your bones from day one, but consider them and think about them regularly, and they will slowly start to take over. **Create that mental environment and step-by-step you will naturally adapt to it**. It would be impossible to do anything else.

Having upgraded your emotions and beliefs, you are on your way to a technologically advanced mind. Up goes your GPP and don't you feel just great as a result? Life is how it was always supposed to be. Nothing in the real world has changed, but it will only be a matter of time before you wiggle your way into better places thanks to your new and improved mind.

Your relationships

Great, so you have a better mind now. But not only does technology exist inside you, it exists across people. You work *together*. You *always* work on things together, and that collaboration can be improved through technology.

The most obvious working relationship is between you and your special someone. At the beginning of the relationship, there was communication. You liked what you heard. Not only the words, but how they were said. Promises were made, affections declared. Maybe outright, maybe subtle and disguised. And it's on! You're a couple. Now what? What's the plan? You got stuff in mind? Does he have the same stuff in mind? Comedian Steve Harvey tells women to ask men right on the first date, "What is your plan?" That way you can gauge if it fits with your plan. That's an interesting communication technology. If it matches, we move on to executing the plan. Maybe you want kids. Maybe you want to just live together and travel. Both scenarios can go well or poorly and the execution of each will depend on the technologies you are using within that relationship. Yes, you are an individual and you determine your own life, but you will be limited by your environment. If your better half doesn't work well with you, there will be problems. Technology is here to help.

The technology of relationships not only applies to you and your special someone, but to you and your kids, you and your boss/employees/colleagues, and you and your own parents. How do you manage those relationships when everybody wants something different? What are you providing and what are you getting back? Just like in any evolutionary exchange, you should be getting back more than you are putting in. But the beauty is that the other person gets more than they are putting in, too. It's trade, man! Everybody's happy.

Relationships are the hardest part of life, but that also means that's where you score the highest. This is one of those subjects you don't learn about in school, but you can study it just like math. Amazon has a special section of books dedicated to relationships, and they're a lot more relevant to your life than math books, let me tell you.

Here's a preview:

How to talk to your kids

Kids and adults live in two different worlds. Having the best experience with kids means talking to them in a way they can relate to, one that pushes the right buttons in those minds of theirs — minds that are looking for independence but still trying to please and navigate a confusing world. I actually have an audio book called *How to Talk to Your Kids* and after I popped it in in the car, it was smooth sailing from there. With a few tips and some perspective on how to change my approach, I got immediate results. How many parents are educating themselves like this these days? More and more. Just a few generations ago it was normal to hit your kids when they were misbehaving. While that may feel natural — I naturally want to smack my kids on occasion — it's a poor investment, and we have better ones these days. Families who learn to communicate well with each other have the same technological advantage that a family with a rabbit net had thousands of years ago. Just like the rabbit net, it can be passed down. Good parents produce kids who become good parents. Get it? Investing through technology, man.

Turn off the "I have to be right" button

As I said, when my wife and I had had enough of all the arguing, we went into couples' therapy. One of the fascinating techniques we learned there was to give up on being right. Fights in a relationship are about emotions and feelings which have been stomped on. There is no "right" to be had in an argument. Sure, one person *is* right and another one is being totally unreasonable, but you're not going to get the hostages out alive if you just start firing shots back. You need tactics and diplomacy and that means letting go of being right. This feels very awkward at times, because it goes against so much of our upbringing and hard-wired tendencies. But if you want to maintain the intimacy in your relationship, it's a much better technique than trying to knock someone down who is already hurt.

Learning to understand the perspective of others

Let's take bosses for a sec. It is human nature to overvalue our own position and undervalue that of someone else. This is why we have wars. I think *you* were out of line but *my* intentions were good, so you are wrong and deserve an ass whooping. And the other side thinks the same. This "taking sides" happens everywhere in all "classes" – rich vs. poor, kids vs. adults, employees vs. management. But of course, all these factions have to work together, and understanding the needs and ways of thinking of the other side will provide more solid ground for cooperation. What does your boss want? What are his goals and what does he want from you? If you can get into that way of thinking, you'll be better able to help. Yes, you will expect the same consideration from him, but one of you has to go first, and your goodwill may go unanswered for a while before being reciprocated. But hey, you're a giver, remember? The same goes for your spouse, your kids, and the people on the other side of the tracks.

Listening skills

It's hard to listen to people . . . they talk so much. Let's look at couples again. There is often a lot of white noise — all the work and emotion — obscuring what's really going on. Usually when we want people to listen to us, it's when we need something and the other side is not all that interested in what we need. There is an unavoidable selfish element in couples, whereas it's more gene-ish with our children. But there is usually a difference between what one side wants and what the other side wants. And in the middle are a lot of poor listening skills. There is much progress to be made when both sides can agree on a form of listening when one side is presenting a need. It is a technique like anything else.

The wanting side needs to clarify and be bold. Don't half-ask for the things that you truly want. Don't beat around the bush, come out and say it. Say it like you deserve it. If you're on the other side, you need to shut up, both verbally and mentally, and really hear the person, without thinking about what it's going to mean to you. Because if your man or woman wants something, damn sure *you're* going to have to pay for it. But don't think about the cost for now, just listen. Feel what the other person feels and then work together on a solution.

Noticing Patterns

Noticing patterns is an important technique. Here's an example: Before going to see my family, I would get stressed out and become this unpleasant person. Sometimes my wife and I would even get into a fight because of it. You would be an idiot not to notice a pattern there. "Every time we are about to do A, B happens." Yet, you miss it, because you want to put it down to other things. By the time you realize you're having an argument, you're already five minutes into it and you don't remember what set it all off. If you can notice patterns and can admit that they are there and have an effect on you, you can take steps ahead of time, or even during, to take control over them. Patterns are another form of software — "if-then-else" statements — and you can re-write them.

All the examples above are just a sampling of the technologies that will be at your disposal if they are practiced. You will be bad at them at first, but you will get better. My relationships are not perfect, but they could be worse. I grew up in a family with constant arguing. I could have the same problem now, but I don't, and it's thanks to developing skills like these. All people want to be in loving, happy relationships, and the only thing stopping them is a lack of technology.

Cultural and digital worlds

Our world is amazing because of the culture and the digital systems we continue to build. Without all that, we are just another animal species fighting and fornicating all day in the middle of nowhere. Success is to be had by those who get the most out of culture and the digital sphere.

Being culturally advanced means using the best cultural technologies at your disposal. In business you have strategies and tools at various levels for various things. For your health, you have diet and fitness programs. For your looks, you have the fashion and beauty industries. For your knowledge resource, you have a dizzying array of education options such as physical universities, online courses, mentoring programs, internships, books galore, and even the good ole internet.

And today, using digital systems is practically mandatory for a fulfilling life. Without them, you're limited in the number of people you can reach, the amount and quality of business you can do, and the number of products you can choose from. Maybe you're not on Facebook, but the company you work for is doing its best to stay competitive, and it's using digital systems to do so. Like it or not, you're on the grid. And with all the benefits it brings, you should like it.

Get on board with cultural and digital technologies. Spend a few minutes here and there reading up on new technologies. What links could I possibly give you that you can't find online in two minutes? OK, go to fastcompany.com, or techcrunch.com, or wired.com. They won't have as broad a definition of technology as I do, but it's a good start. For the softer side of technology, just get into personal development books and programs and even blogs.

Realize that investing in these new technologies will be uncomfortable at first, as any investment is since you're always giving something up. But stick with it and enjoy the benefits those investments give you. It'll be a slight advantage, but that's all you need. Do it over and over and you'll have a new technological life before you know it. Just remember you need to see results in the physical world. Ultimately, we live in a physical world and success in life is a physical thing. Your family will be physical and your contribution to the world will be felt by other human beings. Hug them, pinch them, punch them in the arm — they are physical, and they will tell you if you're succeeding.

6. Don't Look Too Far Ahead

"We kind of had a staircase, and we both knew what step to take next on that staircase."
– King of GPP, Paul McCartney when asked about the secret to The Beatles' success

The Principle

Natural selection has no foresight. It never wanted to create birds, it just kept rewarding those animals that jumped further, and further, and further away from predators until they were flying. The conditions of the next generation are too complicated to predict, so plants and animals can only concentrate on what's going on in their current environment.

Humans, of course, do think about different environments. "What will I be when I grow up?" "Will I be pretty, will I be rich?" All valid questions, but thinking too far ahead can cause problems. The future remains intangible and the here and now is hard enough to demand most of our attention.

The Application

As I've said earlier, evolution progresses in tiny increments. That means you have to take small steps. There are those who say to visualize a big goal in front of you and work toward it. But if you don't have a big goal, consider an alternative: **only look as far as you can see**. Nature is not comfortable with things that are too far away. What works now? Just work on what's in front of you and give it your all. Little by little you build a future.

An example is the Rolling Stones. Keith Richards had a clear goal: to be the best blues band in London. That was it. They accomplished it. All the rest — the other 90% of their career from that point was just taking small steps. Some people have a crystal clear vision of a massive goal and they go get it. Great. Cuba Gooding, Jr. knew he wanted to be part of the Hollywood elite. Big clear goal, and he worked right to it. Not all of us have this. We find ourselves in a certain situation and don't know whether to continue down this path or change for something else. What do you do? You work toward a goal that is crystal clear no matter how small it is.

Yes, you need a big picture and a definition of who you are and what you do. Do you want to climb a ladder in a corporation, start your own business, be an artist, help others directly as a doctor or therapist, be a mother, etc.? From that definition you decide if the job you're working on at the moment fits with that big picture. But we're not looking too far ahead here, and the exact details of where you will end up are better left to fate. All you really want defined is a general level of success. Are you going for a 10 in life or a 7? 7 isn't bad, so no shame in that. It should create a more relaxing life than going for a 10. Hey, nature says get as high a GPR as possible, but we don't have to be total slaves. We can tell her, "This level is good for me."

Taking little steps does not mean you're taking it easy. Life is short and you need to move fast, which means taking little, but frequent steps. The worst times of your life should be the long stretches where nothing changes. Ah! I'm thinking of a few now myself. When you're working every day, banging out your little list, creating the right environment, learning, and compounding it all, you should be changing often. As we've seen, evolution loves change. If you change for the better, you'll be rewarded handsomely.

So, let's recap here. These principles are all taken from the most trusted source in all the world: evolution. They are not going to apply in all cases just like any river will have slight moments where the water is pushed up or to the side and not downstream. But they are playing the odds. There will be enough moments in your life where you will find success in changing quickly or in aiming your arrows into the distant fog. What works, works. So, never let any system stifle your flexibility. If anything, evolution wants you to be flexible, even if its own bottom line is not.

The Evolution of You

So goes the evolution of you. Natural selection is still running the show and, hopefully, this book has shown you how you can take some of its principles and fold them into your life in a very natural way. They should feel natural because they are natural.

We want to understand evolution because we want to do better in life. If we do what nature wants, nature will reward us with a fulfilling life. When we piece it all together, we see that we need to use technology to better accomplish what she wants. This flies in the face of the current thinking that technology and nature are polar opposites, but I hope this book has proven that they are, in fact, related, and that it's given you a new framework for integrating the two.

So what did you do today? Did you improve yourself on some level? You're reading this book, so that's positive! Did you improve someone else on some level? If you did, consider it a win. If you did nothing positive, how does that feel? If you feel like crap, then maybe that's the reason. Instead of thinking you need to treat yourself to some fun and a bag of chips, consider the possibility that you need to do the opposite: You need to work. Maybe you don't really think about your day in those terms, but you should think about what nature wants from you and start doing it. See how it feels, and if it feels good, and it should, then do more of it.

What an amazing book! I sincerely hope you enjoyed it and that you gained some new perspective on life. There's a whirlwind going on around you, but I hope you understand it all a bit better now. For me, this understanding added some well needed distance and objectivity. I don't need to feel so emotional about everything bad that happens. Instead, I feel awe at the beauty of the system that nature has set up. She's been working hard for billions of years, and we're always on the cusp of what she's producing.

Natural selection is the Great River in our world, whether we believe it or not. We might not understand it as precisely as we could, but that is part of the deal too. Understanding is a process. I'm confident that the basics are clear and the little details will only add bits and pieces of clarity and enjoyment. If you go with the flow, you should have a better experience than trying to go against the current. You can do anything you want in this life. That is, you can behave and act as you want, but you want to feel good. That is a human universal. So, you should do those things that make you feel good. I'm confident that nature has set up human life so that doing things that feel good result in positive outcomes for mankind (in the long run, of course). There are still horrible people making a mess of things, but they are slowly becoming obsolete as technology creates an environment in which they can no longer live.

Every new generation is the luckiest generation born. We think that the World War II generation got a raw deal, but only when compared to generations after them. The generations before them got reamed too. We're starting to see more clearly how every generation enjoys the benefits of advanced technology. As confusing as I've found life, I take some comfort in the fact that there is no reason why I should not have a fulfilling, happy life. I understand life now, and every day is like walking through a museum and being amazed at the incredible works of art surrounding me. Life: love it or leave it. And soon you will leave it. So, try to leave it in better shape than when you got here. Not because it's the right thing to do, but because it's the gene-ish thing to do.

Take a while to let this all sink in. I hope it's added something to your life, because that would mean some extra GPP for me. I get higher GPP when you get higher GPP. Get it? Of course you do. If you have any questions or comments just let me know. I'm sure I'm checking email.

So that was it. Am I forgetting anything? Oh yeah, just one more thing. . . .

6 | The Meaning of Life

Welcome to the anti-climax. If you've been paying attention, you might have guessed that there is no meaning to life. Meaning is something our conscious brains thought up trying to force pattern recognition on everything. Since we were created through evolution, there is no one moment when we can say, "This is the first human," just like you can't say, "This is the first salamander." It's a spectrum on which no single point means more or less than the one next to it. Was there a moment when meaning was born in the world? If we think back to a time when the earth was covered in nothing but pond scum, then we can probably agree that the "meaning of life" is a weird concept. At which moment along the evolutionary journey did meaning become a factor? Never, I would venture. Sorry.

You might think that the fact that we dominate the planet indicates that our position is a sort of destiny. Like "OK, it's just pond scum for now, but in a few billion years, humans and their reflective sunglasses will dominate the earth, so it's part of a divine plan." But I might point out that we don't dominate the planet. Yes, we have screwed over a lot of species in the past 1000 years[49], but the rest are doing OK. There are so many ants and mosquitos on the planet that you have to think ahead when planning a picnic! As spring settles in, I just had to tell my kids,

[49] I know. Makes me sad too.

"Only open the windows with the screens on them, because those mosquitos are everywhere." Bacteria, and even plants are doing great now that wild fires are under control. We're doing very well in terms of GPR, but we are still bound to the physical laws of the universe. Even human intelligence, the biggest trophy on our mantel piece, is "just an adaptation": it doesn't give life meaning.

What remains is that technology is *our* adaptation. We may not have meaning, but we have a position to play in this game, and it is a technology game. Now that you know that technology is also sex and love and social interaction, you can fulfill your role as a technician with added brazen. Go out and love and contribute and be one with the rest of the world. That is the recommended position for you these days. Play it well, and you will fulfill the goal that nature had in store for you.

Does it matter that there is no meaning to life? I still love to hold my baby boy in my arms, still love to play soccer with my older boys. I still love to see my wife's smile and feel as in love with her now as ever. I still love to be in nature. I still love blah blah blah. I wouldn't love any of it a bit more if God had made it. If anything, I love it more by the sheer improbability of all of it. Things didn't have to be this way. When God is in charge, we expect it. We expect God to create a world where conscious beings rule – what else was he going to do up there? On top of that, we're disappointed that there is suffering in the world, like he's holding paradise behind his back and won't show it to us until we guess how many fingers he's holding up. And God has an infinite number of fingers. A natural world view shows that suffering is part of the deal, and we are lucky that it's tapering off with time.

I know my purpose is to help human evolution progress. I'm just a drop in the ocean, but it's real. Evolution has taken us from a bunch of groups on the African Plains to an incredible network of people who cherish art, nature, peace, love, and spirituality. Increasingly, our lives are dominated by these themes. My only wish is that I can help to bring these themes to more and more people on ever increasing levels . . . and that I can consistently beat random 12 year olds playing FIFA online (I'm still a human being).

Is there meaning elsewhere in the universe? Because if there is meaning here, then there must be meaning millions of light years away too. We know we are not the center of the universe. If you're out in the country on a clear night, and all of a sudden you see the stars — all of them — you realize that the clouds and daylight have been hiding the fact that the universe is so incredibly big. In the same vein, the clothes we wear and the technology we have hide the fact that we are just apes - apes like the others, with some big differences, but not big enough to change that fact. And just like the apes, we too are animals, set on a course created by natural selection to do nothing but replicate the genes inside us. We at least have creative license to decide how to replicate those genes. And for a long time, we found things like genocide and murder to be good ways to do it. But step by step, adaptation by adaptation, technology by technology, we are moving to a more inclusive strategy that is still gene-ish, but finding more and more rewards in positive sum interactions. The more people, the more interactions. The more interactions, the more technology. The more technology, the more interactions. And so on. . . . Simple formula, no? It's taken a long time to run that program and here we are enjoying the fruits of its labor.

OK, that's really it. I've got nothing else to say. Enough reading. . . . What are you doing tomorrow? Want to grab lunch?

The End

Special Thanks

Thanks to Thomas Arthur Hamill for all your help with this book. Your feedback was the equivalent of finding a pot of gold. That makes you a leprechaun, Tom. . . .

Thanks to Jani Moon for your feedback. Must have sucked to read and critique earlier drafts, but you did it as a good friend would. You're a special person with special talents, and I'm a lucky guy to be on the receiving end.

...You know, my parents didn't even know I wrote a book. I don't expect they will like it either. Over the years, they have both supported me and been a general pain in the ass, so often I feel we're even. In line with the genetic theme of this book, I can say that I get my philosophical side from my dad; and my mother, unbeknownst to her, is actually a pretty good writer. Their consummation was obviously essential to the creation of this book. Thanks for that.

He goes on like he won an Oscar....

Thanks to Rogier Klop for the coaching sessions that kept this train running early on. Thanks to my editor, Coralie Hunter, for polishing this turd into a shiny mass of satisfying human output. Your technical acumen made this book readable and your intellect and understanding made the arguments consistent and tight. And for an insecure first-time author, you were just the right sounding board — not too hard, not too soft.

And thanks to Christine Pride for setting me up with Coralie. Thanks to David Kass for setting me up with Christine. And thanks to dear friend, Julie, for setting me up with David… Networking Success Story #1…See? That's how it works, kids.

Notes

Go on, use this page to write down stuff you want to remember or even your own notes. I paid for the extra page (or maybe you did), so you might as well use it.

Made in the USA
Monee, IL
08 November 2026

67971461R00194